Creating a Better Climate Future

CREATING A BETTER CLIMATE FUTURE

How you can start solving climate change in 5 minutes a day

PHILIP KENT-HUGHES

MINDSTORM
PUBLISHING

Mindstorm Publishing

Editor: Skye Loyd, www.edit-guru.com
Copy Editor and Proofreader: Laura Pasquale, linkedin.com/in/lauraepasqualephd
Cover Art and Book Design: Sarah Lahay, cestbeaudesigns.com

Throughout the book there are links to various websites. I am not associated, nor do I benefit from, any of these organizations.

Any website addresses listed herein are correct at the time of publication. The publisher is not responsible for content of third-party websites. Please take care with your internet security when accessing third party websites.

Second Edition
ISBN: 978-0-6489300-7-5

Cover image source: NASA image by Reto Stöckli

Gratitude

I am grateful to my parents, Nina and Brian, for always believing in me and this project. Without your ongoing support, this book would not have been possible.

I am also grateful to the people who worked so hard to make this book possible: Es Foong for concept development, Skye Loyd for editorial, Laura Pasquale for proofreading, and Sarah Lahay for cover and internal design. Thank you to the review team for your invaluable feedback: Michelle Wang, Saskia Clapton, Amanda Anastasi, Rowan White, Arthur Ha, and Mandy Lupin. Neil Gordon helped bring to life many of the key ideas and Ken Fornari helped create the book title.

I appreciate all your encouragement and support.

Thank you,

Philip

Contents

PART 1

How We Can Solve the Climate Crisis

"Climate change is a threat to human well-being and planetary health. There is a rapidly closing window of opportunity to secure a liveable and sustainable future for all."

—Intergovernmental Panel on Climate Change (IPCC), Sixth Assessment Report (2021)

WE CAN CREATE A BETTER CLIMATE FUTURE!

Are you concerned about the effects of climate change? Year after year, leaders go to climate conferences, and promise to act, yet carbon emissions continue to rise. Does this give you the feeling that there is something wrong with the way things are? Would you like to be empowered to join millions of people already changing the system?

With growing concern, I have witnessed reports of worsening forest fires, heat waves, floods, hurricanes, and other climate catastrophes from around the world. For many years, I went to climate protests and signed petitions, trying to promote change. Then in 2019, there was debate in the media about declaring a 'climate emergency'. As an advisor in emergency management, I was inspired to start writing. At first, it was important to identify the main obstacles preventing climate action and how to help people create system change.

Then I used emergency response principles to develop step-by-step solutions. With simple procedures, even a daunting task can be completed. Life is busy, so I've organized the areas of the response into manageable sections, such as food, transport, energy, production, and consumption.

Everyone has their own situation; some have financial constraints, while other people have different challenges. I've provided a wide range of options to be inclusive. Choose the actions that work best for *you*, tackling the easy ones, then moving to others that are more involved. Many actions take only a few minutes and can be included into a daily routine.

In this book, you'll discover checklists focused on reducing emissions in individual households, and other activities designed to influence corporations to apply sensible and sustainable solutions. There are real-life examples of how I applied the options, where I had difficulties, and what I did to address them. You will have the choice to take action individually, with your friends and family, or even with people in your town, city, country, or all over the world. There will be opportunities to find new community and make new friends.

We still have time to turn things around. We can worry less about the future by knowing that we are doing our part to make the world a better place. When we join the rising tide of people acting on the climate crisis, we will generate an unstoppable force that will change the direction of corporations, industries, and even economies.

What is the *real* problem?

Climate change isn't the problem; instead, it's the consequence of ignoring the science. Or as UN Secretary-General António Guterres noted, "Climate change is the defining issue of our time" and "Scientists have been telling us for decades. Repeatedly. Far too many leaders have refused to listen."[1]

Like crime scene detectives, to find the source of a problem we need to review the facts.

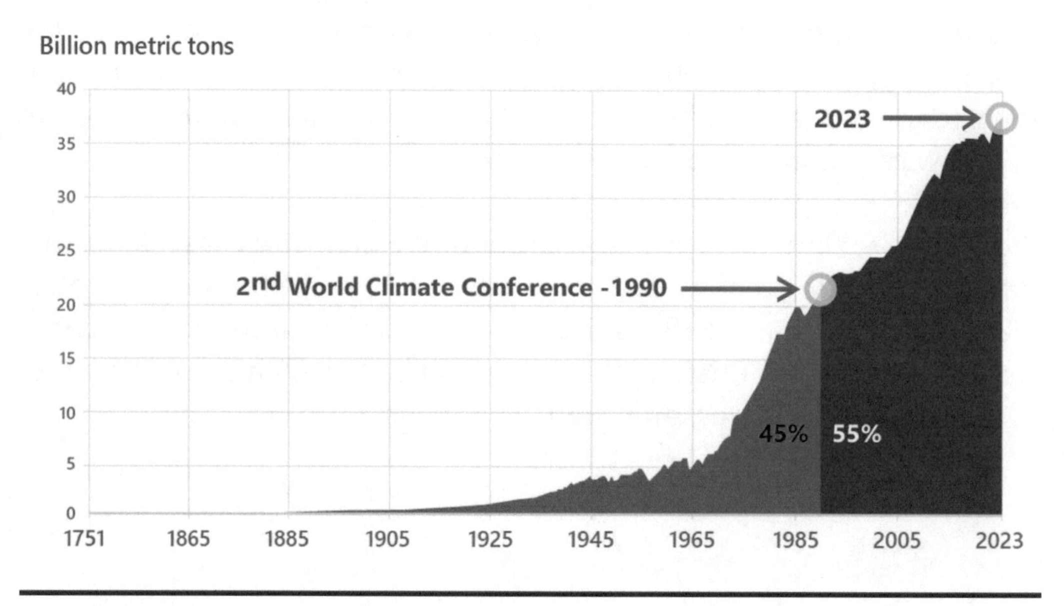

Source: International Energy Agency (IEA) CO2 Emissions in 2022

The international community agreed to act on climate change in 1990, but instead of falling, emissions grew even more rapidly than before (shown in the graph above).[2] Thirty years of failure cannot be an accident. Why isn't change occurring more rapidly?

The real problem is not emissions. **The real problem is there are obstacles to positive change.**

As an emergency management advisor, I've helped large organizations prepare for incidents that may threaten people's lives. Even after writing a new plan and training their team, sometimes there were problems preventing these organizations from responding effectively to an incident. However, I found that if we ask the right questions, we can find out who or what is responsible for the obstacles. Then we can deal with them, and even prevent other problems from happening in the future. To solve the climate crisis, the essential first step is to correctly identify the obstacles which are preventing meaningful change. The main purpose of this book is to identify these obstacles and provide solutions to overcome them.

What are the obstacles?

Investigators found that oil and gas corporations spread misinformation about climate change that their own scientists knew to be untrue.[3] They also worked in secret to defeat the United Nations' attempts to reduce emissions.[4] Other investigations found that oil and gas corporations spent more than $1 billion on climate-related lobbying of politicians and political parties soon after the 2015 Paris conference.[5] It's no wonder traditional methods of creating positive change have failed when opposed by this vast wall of money that corrupts and undermines democracy. (For more details, see Appendix A: "Oil Companies Knew.")

Even considering these revelations, it would be naive to blame the oil and gas industry for everything, because most other industries have done little or nothing to reduce emissions. Science has shown that all sectors have options to halve emissions by 2030, so why aren't the solutions being implemented?[6]

Nicholas Stern, Professor at the London School of Economics, wrote a report on climate change for the UK government. He explained, "Climate change is the greatest market failure the world has seen,"[7] because "those who damage others by emitting greenhouse gases generally do not pay."[8]

Why has the free-market economy failed so badly? One of the main objectives of the free market is profit. Annual profits can decide executive bonuses, dividends to shareholders, and influence the share price. Because of this, many companies often focus on making quick profits instead of investing money and other resources into reducing emissions. This focus on annual profits can also lead to corporate leaders doing nothing about their emissions and passing the consequences on to the next generation.

Some people say that we have a choice between looking after the economy and jobs or the environment. However, this is a false choice—because we can have *both*. Researchers found that companies that include environmental, social, and corporate governance (ESG) into their growth strategy can outperform their competitors.[9] When we abandon false ideas about the free market economy, then we can solve climate change.

We can have a healthy economy, jobs, and emissions reductions and still create a better world for present and future generations. Nicholas Stern said of the response to climate change, "If we get this right, it will be more powerful than the industrial revolution."[10]

How can we overcome the obstacles?

Most corporations, industries, and governments are doing too little, too late to reduce emissions. It is up to us to act now, and we should not ask for permission or wait anymore. We can guide a new direction for corporations, industry, and even entire economies by creating system change. We can keep the benefits of the free market, such as freedom of choice, competition, and innovation, while creating sensible objectives for the economy other than just profit and growth.

History shows us that when large numbers of people have stood up for an important issue, then change has followed. There have been many important social movements that achieved success, such as voting rights for women, civil rights in the US, the anti-apartheid movement in South Africa, ousting dictators, and marriage equality, to name just a few. Through people power and unified actions, we can change the direction of corporations and create system change that will lead to a better world.

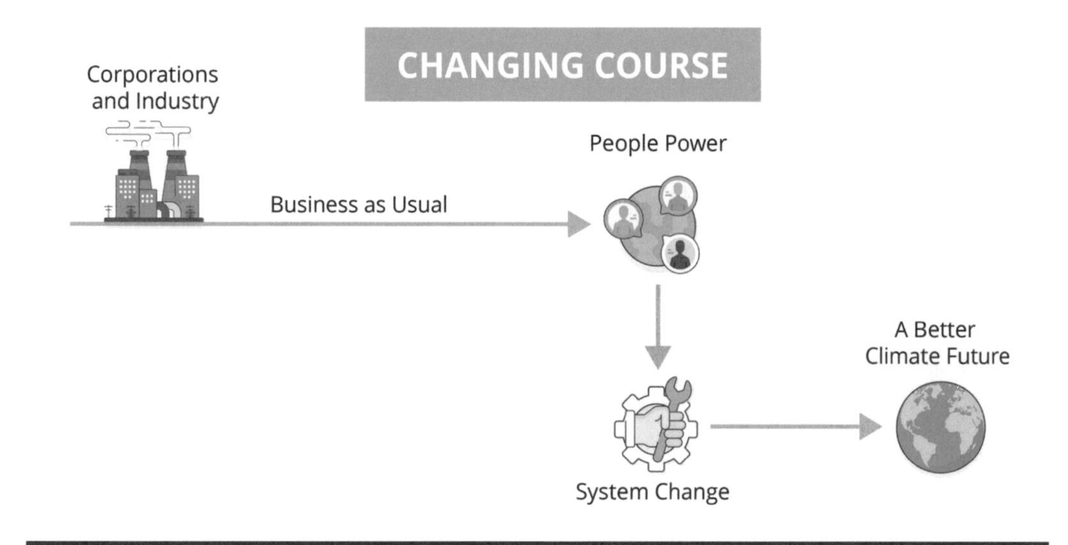

Some people have said we need a revolution to tear the free market down. But what if there was a way of solving climate change by using the rules of the economy to change itself?

How do we create system change?

Corporations have used advertising to tell us what products and services to buy. Some politicians have fueled culture wars to distract us and make us afraid. This is, in part, so we won't hold them accountable for failing to do their jobs, to serve the people instead of the corporations that give them money. It's time we turned the tables on both corporations and the politicians. We have been told what to do and think for too long, it's time to reclaim our power. It's time we tell them what to do to serve our needs. **We the people, can step up to provide the leadership that has been missing. We can reverse "business as usual".**

BUSINESS AS USUAL

Corporations and Industry

Tell us what to buy

PEOPLE POWER

We, the people

Tell Corporations and Industry what to make and how

How can we use the economic system to change itself?

So far, some of the main ways to take climate action have been:

- Protest and lobby politicians to make regulations.
- Take direct action to physically stop fossil fuel corporations.
- Use the legal system, to impose rulings to stop oil and gas expansion.

These remain essential tactics to push for positive change, and people should support them as much as possible. But there's also another way we can make a better future and change the system; it's quite simple and accessible to many people. There are two main features of the free-market economy that can be used as opportunities to make positive climate action:

- **The profit motive and self-interest:** Corporate profits can affect dividends, share price, and executive bonuses.
- **The freedom of choice:** The producer has the freedom to choose what to make, and the consumer can choose what to buy.

Corporations spend a lot of money on advertising promoting products or services to boost sales so they can make profits. In some cases, some have spent years building a popular corporate identity and attractive brand awareness. But what if all of this were at risk? Corporations and entire industries have polluted the atmosphere on a massive scale. Now, we the people can actively use our freedom to choose not to buy from these organizations. We need to recognize our power as a group and take unified actions. If customers decide not to buy products or services from corporations that refuse to change, then this will make their profits go down—and *that* will get their attention.

Now, what if I said that it's happening already, and that all we need to do is join in? Researchers in the US, found that 20% of people said they have punished companies that are opposing steps to reduce global warming by not buying their products.[11] They also found that 35% of those asked intended to punish companies in the following year (see the following chart).[12] People also rewarded companies that were reducing their emissions by buying their products; in 2022, this group included 26% of research participants, and this was expected to rise to 34% the next year.[13] Ethical shopping can be an active expression of your beliefs and a way of living a lifestyle with a clear conscience.

People are punishing or rewarding companies based on their action to limit global warming.

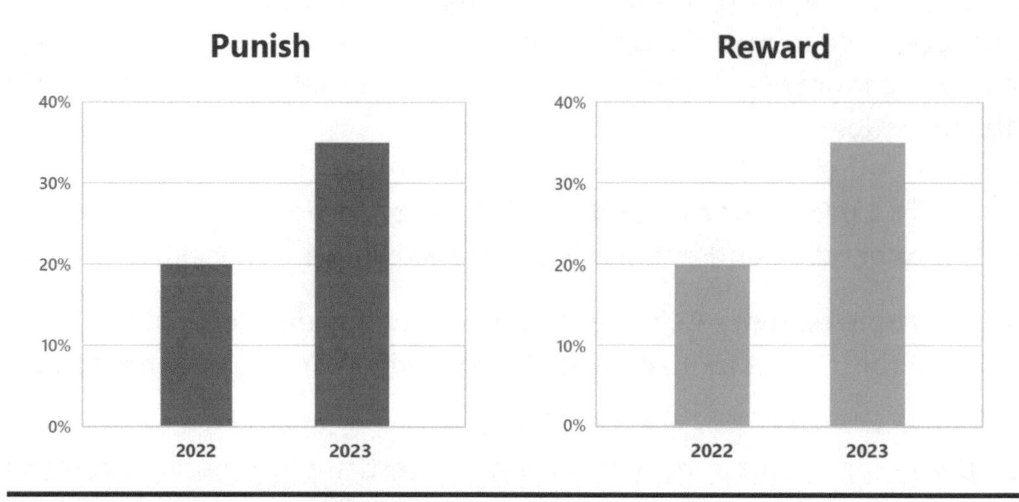

Source: Yale Program on Climate Change Communication

Imagine what could happen if we coordinated our efforts and took unified actions together. The figures for people punishing unethical corporations and rewarding good ones could easily increase to more than 50% or even 60%. We can reduce corporate profits with our purchase decisions until they reduce their emissions. When profits fall, then our voices will be heard in boardrooms across the globe. The risk for corporations is that unless they change, they may lose customers and never get them back.

Amplifying self-system change

To super-charge consumer campaigns, we can use additional strategies. Research has shown that corporations are much more likely to agree to the demands of social movements if they also suffer bad publicity.[14] This can include viral posts on social media, unhappy and complaining customers, protests outside stores, and other forms of negative media attention. Bad publicity can create a crisis for corporations by compromising their reputation and challenging their social license to operate. We can name and shame them into making positive change to reduce emissions. Some of the corporate policy changes driven by consumer campaigns include:[15]

- Nike improving working conditions of the people making their brands.
- SeaWorld ending its captive orca breeding program.
- Nestle committing to reduce deforestation by not purchasing palm oil.
- Kimberly-Clark creating new paper-buying policies to reduce deforestation.
- Zara clothing stores eliminating items made with fur from one thousand stores.

Social media platforms provide an unprecedented level of intensity and visibility for potential new campaigns on climate change. A campaign started by a small group of people could quickly cascade and be supported by thousands or millions of people in hours or just a few days.

Consider this scenario: a group decides to take unified action and create an online petition demanding a corporate retailer reduce their emissions. This is signed by 100,000 people and then delivered to the corporate headquarters. If they don't respond, then another unified action could create a campaign on social media and rally people to protest at dozens or hundreds of their retail outlets. People could be asked to reconsider their ethical purchasing decisions for this high-emissions polluter. This type of unified action could get significant national media attention. The opportunities to take unified actions with other people will be further explained throughout the book.

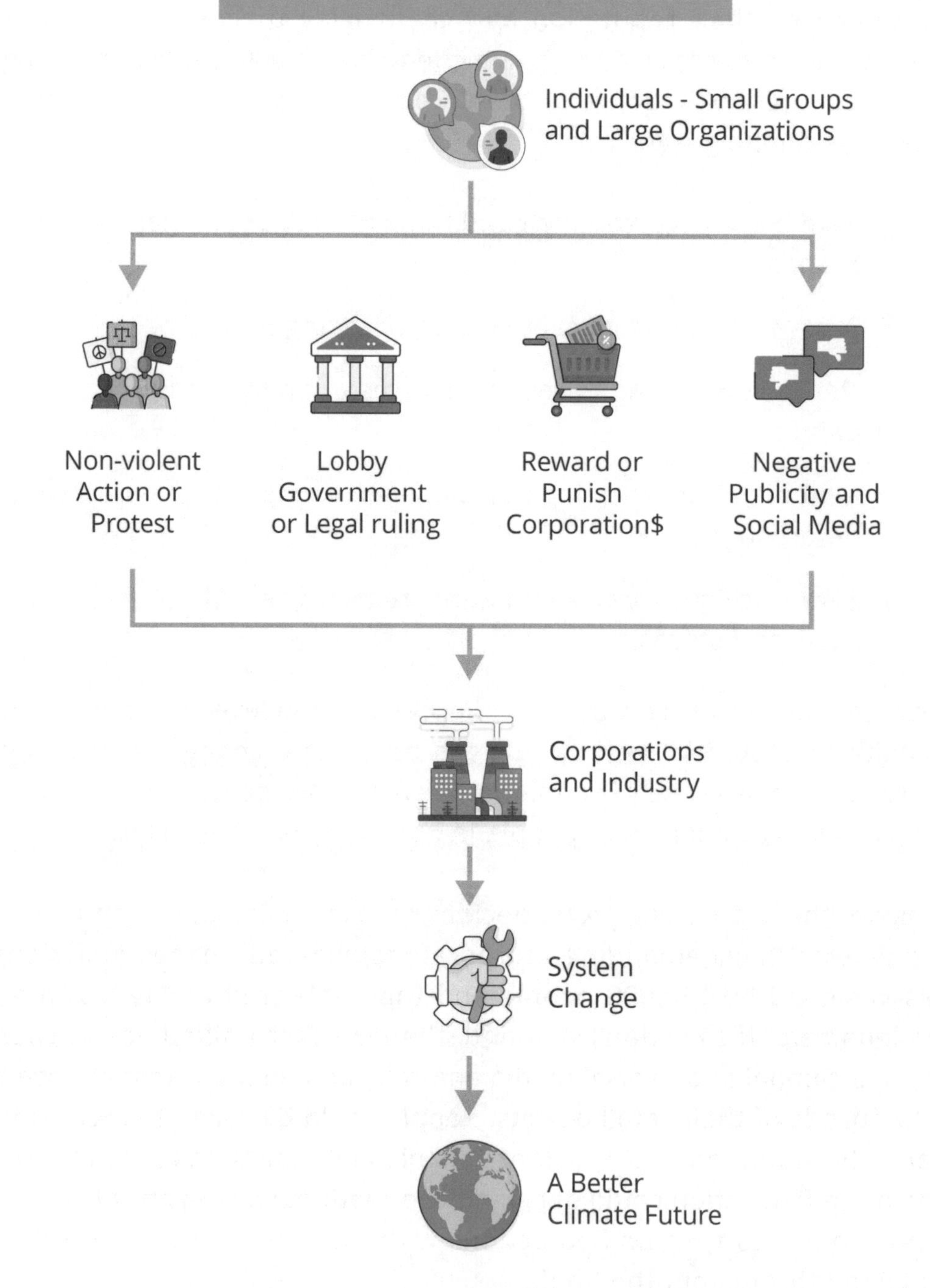
HOW WE CAN MAKE CHANGE
Individuals - Small Groups and Large Organizations
Non-violent Action or Protest
Lobby Government or Legal ruling
Reward or Punish Corporation$
Negative Publicity and Social Media
Corporations and Industry
System Change
A Better Climate Future

As noted, in a free market, corporations can decide what to produce, but consumers also have the freedom to choose what they buy. Encouraging people to think about what they buy and who they choose to buy from is like encouraging people to vote for a particular candidate in an election. Unified actions apply the principles of both democracy and the free market. The aim is not to put any corporation out of business. The aim is to influence them to reduce their emissions the way they should have started to do decades ago. We can also use these methods to reduce inequality, reduce gender discrimination, and challenge other social issues.

We must do this

If Earth had a main control room, its emergency lights would be flashing, and sirens would be wailing. Immediate action is more important now than ever before because burning fossil fuels is increasing the impacts on people and ecosystems all over the world.

Projected Rise in Global Temperatures

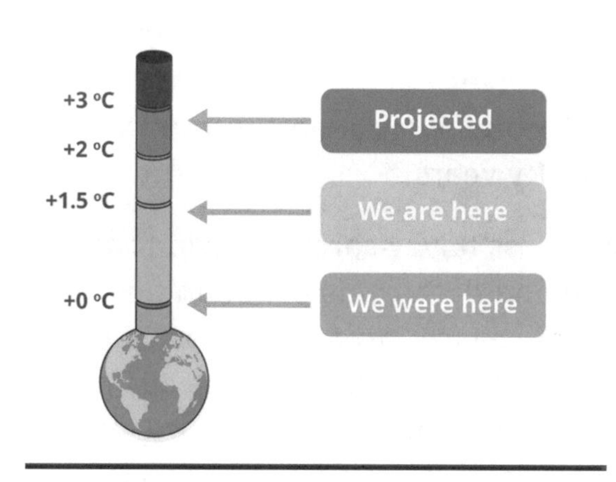

Source: Climate Action Tracker

Global government policies currently in place are projected to result in a temperature rise of 2.7°C and possibly as high as 3.4°C in this century.[16] It is possible that with the pledges and targets made by governments, it might not become so warm. But I don't believe in relying on the promises of political leaders because they have failed to take action for decades.

The Intergovernmental Panel on Climate Change (IPCC) was created by the United Nations to provide policymakers with regular scientific assessments on climate change. The latest report involved more than 700 scientists of 90 different nationalities.[17] The IPCC have warned that "every extra bit of warming matters, especially since warming of 1.5°C or higher increases the risk associated with long-lasting or irreversible changes."[18] Action is essential because burning fossil fuels will worsen the impacts on people and ecosystems all over the world.

- **Rising sea levels:** Entire communities have already been lost to sea level rise.[19] In the coming decades, entire island nations could be inundated, and face forced relocation.[20] Every year, more people will lose their homes, and based on current projections, 360 million people will be threatened by annual flood events by 2100 in a +2°C scenario.[21]

- **Displacement:** As temperatures rise, many people will be living in areas that will become too hot for humans. This will affect an estimated 1.5 billion people by 2070.[22] What happens if tens or hundreds of millions of climate refugees go on the move? Where will they go?

- **Disease:** Researchers found mosquitoes are already spreading beyond their current habitats because of global warming.[23] This will increase the risk of dengue fever, zika, and West Nile virus for about 1 billion people in Europe, Russia, northern Asia, and North America over the next fifty years.[24]

- **Ecosystem destruction:** Wildfires have scorched regions all over the world, killing wildlife and destroying communities. In the polar regions, the loss of sea ice threatens polar bears, seals, birds, fish, and whales.[25] Ocean heatwaves are causing the mass die-off of coral reefs, which are nurseries for fish and many other species.[26] If the global temperature goes to 2°C, then it is expected that 99% of all coral reefs will die.[27]

To learn more about the impacts, and feedback loops that can escalate the situation, see Appendix B, "Impacts and Crisis Escalation."

We can do this

Now is not the time to be a spectator and watch the world burn.

Imagine the people you care most about. Now consider: what positive action are you willing to take to protect yourself and give them a healthy future?

Although it's easy to become discouraged about how bad things are now, positive change never comes from submission, resignation, or despair. The promise of this book is to show how we can be empowered to join in and influence urgent action on climate change.

In fact, people all over the world are already making choices to protect our future. Each time someone stops supporting an unethical corporation and chooses one that actively reduces emissions, they have 'voted' in favor of a new system, one that cares for people and nature. The choices we make, as part of a wider campaign for system change will be the breakthrough we need.

World-renowned anthropologist Jane Goodall describes the effect we can have: "Every single day, each of us makes some impact on the planet and we can choose what sort of impact we make. It is the cumulative effect of millions or billions of ethical choices regarding what we buy that will move us toward a better world."[28]

We already have many of the solutions, but to make this happen, we need as many people and organizations as possible to be involved to accelerate change. We can create community and rediscover our power to decide how the system should work for us all.

The past teaches us that those who show up and take action decide the future. When the history books tell the story of our time, I believe that because of what we do now, they will say that millions of people around the world united together. And that every day more and more people took action, until there was an overwhelming tide of change, reducing emissions and protecting our beautiful home.

You can be part of this defining moment that changes the course of history. Together, we can create a more positive future for all life on Earth.

PART 2

A New Emergency Plan to Solve Climate Change

"The climate we experience in the future depends on our decisions now."

"Our choices will reverberate for hundreds, even thousands, of years."

—Intergovernmental Panel on Climate Change (IPCC), Sixth Assessment Report (2021)

HOW TO SUCCESSFULLY USE THIS BOOK

Choose your own adventure!

There is a variety of options for taking action. In addition to reducing your own emissions, you can also spread positive success stories. You might be happy using social media or speaking face-to-face with friends, while others may prefer to work on their own. You could also participate in influencing government to act, signing petitions, or protesting polluting corporations. There are many ways to help create change, so choose the ones that suit your situation. There will be some easy actions to start with, and I'll also give examples of how I applied them. While I have written this book with individuals and small groups in mind, many of the steps can also apply to businesses and organizations.

Most global greenhouse gas emissions are from these main areas: food, energy, transportation, as well as production and consumption. So, to make it easy, the implementation part of the book is divided into these sections. Each of these includes an introduction that covers the situation and then five action methods. Each action method has an icon, a color, and a banner, as shown below.

Action Methods

PERSONAL ACTIONS	REDUCING YOUR OWN GREENHOUSE GAS EMISSIONS
COMMUNICATE	SHARING INFORMATION OR YOUR EXPERIENCES
CONNECT	CONNECTING FAMILY AND FRIENDS, OR JOINING GROUPS
INFLUENCE	ENCOURAGING POSITIVE CHANGE
COMPEL	HELPING THOSE RELUCTANT TO CHANGE

The action methods include an easy-to-follow, step-by-step checklist. While working through each section, I suggest going through the checklists and attempting as many actions as possible. You can review the personal actions to reduce your own emissions, then choose any of the other action methods that work best for you. All actions are offered with the understanding that the only thing being asked is that we do the best we can, considering our own circumstances.

Choose your own level of involvement

You can focus on actions you enjoy, the ones you are good at, or those where you have a specialized skill. The choices might align with one or more of the different levels of involvement, which can include:

- **Lifestyle changer:** Reduce your own emissions and be an example of how this is possible for others. You could choose to share your success in person or online.
- **Ethical campaigner:** Change your purchasing decisions on your own, with your group or with an environmental organization.
- **Quiet campaigner:** Take part in protests marches, sign petitions, or volunteer in a support role at an environmental group in your spare time.
- **Climate circle organizer:** Create a group to share success and encourage each other. This is explained in more detail in Step 5, "Create unstoppable momentum."
- **Online influencer:** Send messages and spread ideas and information through social media.
- **In-person campaigner:** Volunteer in a more active role at an environmental group in your spare time. This could include lobbying politicians and protesting at corporate headquarters

or retail sites. These actions should *not* involve hazardous activities or breaking the law.

- **Nonviolent direct action:** Join an existing environmental group and take part in actions to raise awareness, put pressure on corporations, or stop the expansion of fossil fuels. An established organization can help prepare you with training, experienced leaders, provide legal and other support. You should be mindful of the consequences of any hazards or breaking the law.

What you choose now may also change over time. You can also try something new. For example, I joined a local environmental group and found a great community. I took part in lobbying politicians to pass new climate legislation and campaigns to influence corporations to reduce their emissions.

The Climate Action website

I've created a website that supports the aim of this book. It provides:

- Information about unified actions nationally, regionally, or globally, so we can all create positive system change.
- A free *Climate Action Guide* for download to help reduce emissions.
- A calendar of events and actions throughout the year.
- A blog of positive emissions reductions news and reports from trusted sources.
- Additional ways to take action with regularly updated information.

Access the website at **www.climate-action.org**.

Unified Action

I will post information from as many environmental groups as possible about actions they have planned, so we can get people all over the world participating at the same time. In this way we can amplify our influence on corporations and industries to reduce their emissions. Actions can include:

- Researching a corporation to find out about their actions to reduce emissions and what their objectives are.
- Signing an online petition demanding a corporation reduce their emissions.
- Deciding about ethical purchasing decisions and whether to pause buying the products or services from a high-polluting corporation.
- Messaging across different social media platforms about the failure of a corporation to reduce their emissions.
- Encouraging people on social media to join in and take action.
- Protesting at retail outlets or corporate headquarters.

When I started, it was important to choose a level of involvement that I was comfortable with. Later, I joined in with new actions, one at a time. As many actions as possible will be listed on the website. You can also sign up for an email subscription to be notified once a month about various actions.

The Climate Action Calendar

The calendar on the website lists a range of activities each month, as well as major international days that are related to climate action and the environment. Examples include World Wildlife Day, Earth Hour, Earth Day, World Environment Day, Plastic-Free July, World Car-Free Day, and others.

Free Climate Action Guide

While working through the emissions reduction checklists, it may be useful to plan. Especially any items that might be a work in progress. To help make this easier, I have created a *Climate Action Guide*. Use it to decide what to do, who will do it and when it will be done. Then mark off actions as they are completed. The guide also has more detailed information to help with each step. It can be downloaded **www.climate-action.org**.

Climate Action Guide: Example

Personal action

Description of the Action	What An action summary	Who will do it	When to do it	Status completed
Check freshness Once or twice a week, go through the fridge and check meat, dairy, fruit, and vegetables for dates and freshness. Then prioritize eating food that could soon go bad. Use mature fruit and vegetables in smoothies or juices.	Check fridge Check cupboards Check freezer	Person A	Every time before shopping Saturday	☑
Planning saves money Taking a few minutes to plan what you are going to eat for the week can make the process easier. Deciding what recipes to use will help to work out the ingredients. Check what you have on hand, then make a shopping list. Also, avoid impulse buys.	Decide on main meals for the week Review recipes and check what you need to buy and make a list	Everyone Person B	Every time before shopping	☑
Not typical is beautiful We can help retailers reduce food waste by buying oddly shaped fruit and vegetables.	Buy fruit and vegetables which are not perfect	Person B	Every time shopping	☑
Make a date The date you need to know is the "use-by" date, which is the last date recommended for the use of the product.[11] Use this to check freshness.	Look for the "use-by" date for freshness	Person A	When checking food	☑

CREATING A BETTER CLIMATE FUTURE

CLIMATE ACTION GUIDE

PHILIP KENT-HUGHES

STEP 1: BECOME A CLIMATE HERO

Many people look up to and admire great leaders, people who have risked their lives to save others, those who have succeeded in principled achievements, or even fictional characters. We are both the main character and author of our own story. So, let's recognize that and appreciate that power. If we decide to take action and strive to create a better world, then we can be the hero in our own story by playing our part in changing the course of history.

Be a change agent

We can respond to climate change by recrafting our lifestyles. We can achieve this by altering our existing habits and behavior patterns. Change is not always easy, but it doesn't have to be as hard as we sometimes make it. The book Atomic Habits by James Clear helped me understand that by making many small changes that gradually add up, we can make a big difference in our lives.

Create a new identity

Atomic Habits also suggests that instead of focusing only on the outcome, a good starting point is to change our belief about who we are. Don't worry, it's not too drastic. I decided that part of my new identity would be the statement "I am living a sustainable, zero emissions lifestyle." Even though this is something I am aiming for, this statement is written in the present tense. I find that this prompts my brain to unconsciously work towards it. I read it frequently, and I find it helps to reinforce my thinking and motivation.

PERSONAL ACTION CREATE A NEW IDENTITY

Your new identity

Write out a belief statement about your own new identity and who you want to become.

Add the identity statement to your Climate Action Guide.

Creating a positive future through our choices

STEP 2: ASSEMBLE YOUR TEAM

Your team can be your household or a small group of people who work together on reducing emissions and taking action. Either way, it's important to include any household members at the beginning of the process. It might be the case that the people you share a home with have different ideas on climate change. Before discussing this with them, think about their values: what do they prioritize in their lives? Instead of just trying to convince them about the "right" actions, speaking to people in terms of what is important to them can make a big difference. Depending on the person, consider discussing how reducing emissions will help do the following:

- Reduce the worsening impacts of natural disasters such as fire, drought, and hurricanes.
- Protect natural ecosystems upon which we and our children depend.
- Reduce the amount of energy used, thereby lowering costs, and saving money.
- Look after the Earth, as instructed by nearly every religious faith.
- Bring people together in fun activities, such as growing a small vegetable garden together.
- Listen to what they have to say and look for areas of agreement and shared interests and values.

Choose an option that most aligns with their values and respects the way they view the world.

What I did

I had a talk with my father about climate change and writing this book. We discussed how this was an opportunity to reduce our emissions and contribute to positive change. We also talked about how in some areas, there was potential to save money as well. We both agreed to continue the conversation. When I was young, my family would watch David Attenborough's documentaries about cheetahs roaming the plains of Africa and the wonders of the natural world. So, when I suggested watching a documentary by David Attenborough called *Climate Change—The Facts*, my father readily agreed. Finding reputable information presented by a person we both respected made it easier to continue the discussion.

PERSONAL ACTION SELECT YOUR TEAM

Select your team.

If living with other people, then these can be the whole team or the foundation group. Then consider whether to include others, such as family, or friends.

Discuss climate change impacts and potential positive action

Consider having conversations with each person individually, or possibly as a group if that would work better. Explain your concerns about the impact of climate change and your interest in finding out about reducing household emissions. No need to cover everything in one discussion; it may take several conversations. Consider identifying some benefits in reducing emissions for the household by reviewing Part 3, "Implementing the Actions."

Add this to your Climate Action Guide.

STEP 3: CREATE A BETTER VISION FOR THE FUTURE

The climate crisis causes tension between the world we live in now and the one we would like to create. But how do we get through that tension? Before a positive outcome can take place, an idea must happen first, then a decision to act, followed by the action itself. After we decide on a clear vision of the future we want to create, we can turn it into reality.

A few years ago, I volunteered with a local environmental group to help write a strategy for action on climate change. We were asked to imagine a future society and picture in our minds what we wanted the world to be like. We placed "2030" in the middle of a large sheet of paper and then wrote all our ideas around it. Some of these included a stable temperature favorable to life, widespread renewable energy, zero-emissions transportation, products made sustainably, while protecting ecosystems and biodiversity. We all have a role in imagining and building the world we want to live in. Creating a vision of the future can help drive your commitment to creating positive change.

PERSONAL ACTION CREATE YOUR VISION FOR THE FUTURE

Create a vision for a better future

This could be a group activity with your team or done individually. Close your eyes and imagine the kind of future you want to help create. Then make a list of these ideas.

Create vision and action boards

A vision board is a representation of ideas for the future in the previous step using pictures, words, or symbols. An action board is a representation of what can be done to make the vision a reality. You can get some inspiration for the action board by reviewing the sections in Part 3, "Implementing the Actions." This could be a group or individual activity. Use a good mix of actions which can be achieved easily or soon. Consider including a few items on the action board that are either difficult or currently out of reach but attainable in the long term.
For example, I can't afford an electric car right now, but I intend to buy one as soon as I can.

The vision and action boards don't have to include everything and can be updated by adding new items or marking completed items. Get pictures from magazines or online and include the identity statement from Step 1 or develop a team statement. Making or printing vision and action boards and placing them in a visible location can be helpful. Put them in a bedroom or communal space, like a kitchen or inside a bathroom door, or keep them on a smart phone as a wallpaper.

Add this vision statement to your Climate Action Guide.

I've made some basic examples using only a few pictures and I've also included the identity statement.

Example Vision Board

VISION

CREATING A BETTER CLIMATE FUTURE

Example Action Board

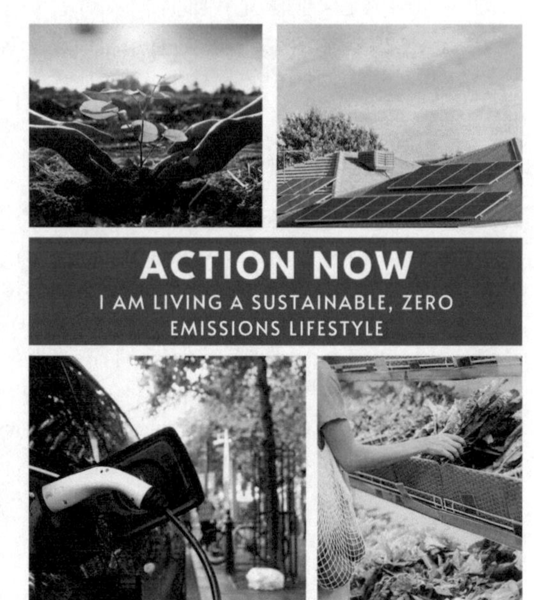

Look at your vision and action boards on a regular basis. This is especially helpful when having difficulty or to fuel your motivation. When looking at a vision board, imagine the feeling of living in a healthier and sustainable world. Do you feel happier and more optimistic?

STEP 4: SET YOUR OBJECTIVE

If we don't know where we want to go, it's difficult to plan how to get there. The aim of this section is to help decide what to achieve when reducing emissions. According to the Intergovernmental Panel on Climate Change (IPCC), we need to reduce our carbon emissions by 45% by 2030 if we want to avoid some of the worst impacts.[29] To achieve this, we need to phase out the use of fossil fuels such as coal, oil, and gas as quickly as possible. This may seem like an ambitious objective; however, we face a crisis like no other in human history. For corporations, industries and government we are using a 45% emissions reduction by 2030 as the overall objective.

Personal or household objective

The next thing to decide is a personal emissions reduction objective. A person could say "I intend to reduce my emissions." The problem with this statement is that it is unclear. With a vague statement, measuring progress or even knowing when the objective has been achieved is difficult. I found that using the SMART criteria is a helpful way to write an objective. I've described these criteria in the following table:

SPECIFIC	Write a defined goal.
MEASURABLE	Can the outcome be measured?
ASSIGNABLE	Who will do it? Collaborate and get agreement with others
REALISTIC	Is it achievable, given available resources, and relevant?
TIME FRAME	When will you start? When will the objective be completed?

What I did

Things don't always go according to plan. When I started writing this book, I was renting a small apartment by myself. This meant the decision-making process about changing my life to reduce emissions was simple. Then my life changed dramatically. I was told that my mother had been diagnosed with terminal cancer. That was a terrible shock, and I decided to move back home to spend time with my mother and help my father look after her. A few months later, my mother passed away, and it took me a while to recover. I stayed living at home, and after some time, I continued writing this book. When it was completed, I had conversations with my father about applying the actions. We decided on the following objective, which we put into the SMART format:

SPECIFIC	Go through the actions in the book and complete as many as possible.
MEASURABLE	We will make a note of all the actions we complete, which ones do not apply to us, and which actions we will need to work on in the future.
ASSIGNABLE	We will work on reducing emissions together.
REALISTIC	The objective and time frame are both feasible.
TIME FRAME	Over the next 12 months.

Write an overall emissions reduction objective using the SMART criteria template. Consider including the people in your household in this process. To begin with, this can be as simple as going through the actions in this book and completing as many as possible. Alternatively, you can calculate your household emissions and then set an objective to reduce them by a specific amount. This is detailed in the carbon footprint process covered in Step 7.

PERSONAL ACTION DECIDE ON AN EMISSIONS REDUCTION OBJECTIVE

You can write an objective using the SMART criteria outlined below:

SPECIFIC	
MEASURABLE	
ASSIGNABLE	
REALISTIC	
TIME FRAME	

Add this SMART goal to your Climate Action Guide.

The benefit of having a clear objective is that you can continuously work towards it with small or large actions on a regular basis until it's achieved. This is not like a game where you might win or lose. You will win if you continue to reduce your emissions and never give up. Each time you complete an action section, celebrate your success: have a party or treat yourself to something special. Also, let people know about it—you can inspire others!

STEP 5: CREATE UNSTOPPABLE MOMENTUM!

How you can help supercharge climate action

When we significantly increase the number of people engaging in climate action, then positive change will surge forward like a large wave. Research has shown that when 25% of a group agree on a course of action and are coordinated, then a critical mass can be achieved.[30] A tipping point is created, and this can overturn established social conventions or business as usual. A global poll of 50 countries found that 64% of people believe that "climate change is a global emergency."[31] **So we don't have to convince people that action is necessary; all we need to do is help them believe that they can create meaningful change and make it easy for them to get involved.**

While progressing through the activities in the book, you can communicate how you are making positive change. This will demonstrate that it can be easy and rewarding to take part in climate action. As we are doing this, if we each encourage 1 person to act and they also encourage 1 person, then after 10 steps there will be 10 new people taking action. That is pretty good—and already 10 times better than 1. However, if we tell 2 people and they each encourage 2 people to take action, then after 10 steps there will be 1,024 new people taking part. If we encourage 3 people to act, then after 10 steps the number increases to 59,049! You can see the difference in the chart below. This is how a social cascade can be set in motion.

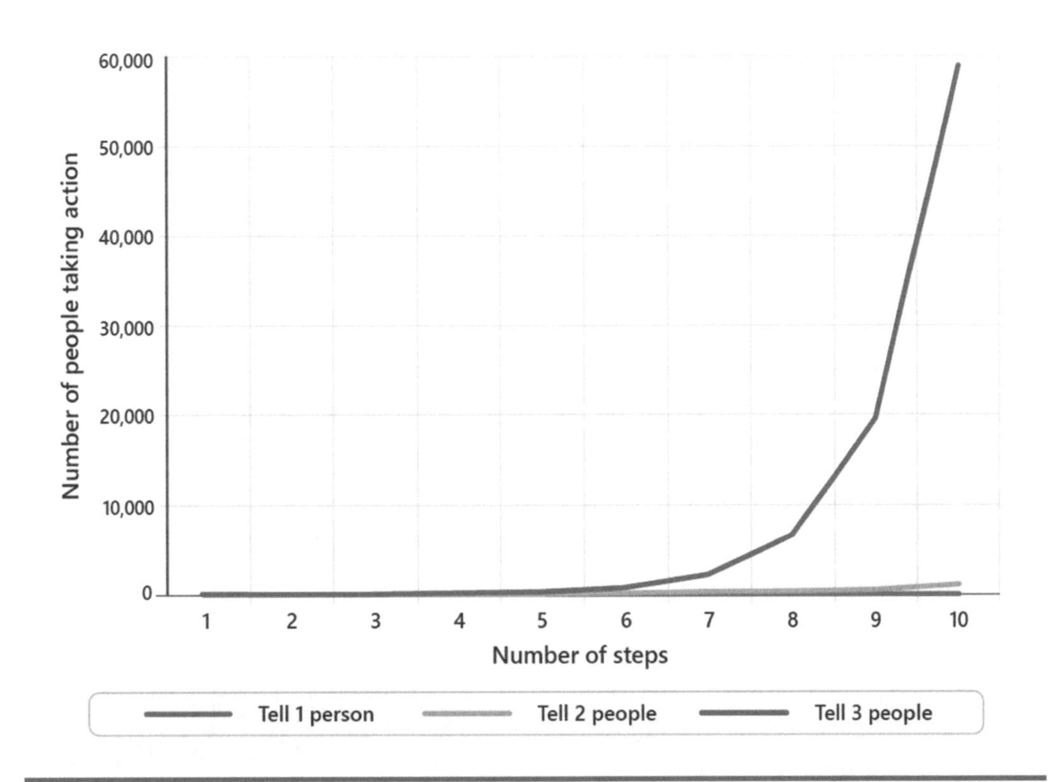

In real life, this happened when the idea for an international climate strike was inspired at the Global Youth Summit in 2015.[32] Support grew when activist Greta Thunberg began striking outside the Swedish Parliament. People started telling and encouraging others through social media. Rapid growth was soon achieved, and millions of school children began protesting globally. Through expanding networks, the climate movement has grown to millions of people in more than 1,600 cities in nearly every country around the world.[33]

Talk to 3 people in 1 year?

Could you talk about climate action with 3 people in the next 12 months? You could start the conversation about the activities you've completed. For example, I've mentioned to friends how my father and I have made a vegetable garden, and how tasty and healthy our food is now. Also, I sometimes let friends know when I've found a new sustainable product. If you post about your successes online and a friend likes or comments about it, you could continue the conversation with them.

People you choose might be friends from school, work, or a sports team, from those involved in a hobby, or from a faith community. They could be people from the local neighborhood, your town, or across the city. It would also be good to include anyone you might know in another state, region, or even a different country. This would help connect people all over the world.

CONNECT **CONNECTION LIST**

Start with thinking of three people to encourage and write their names down in the table below.

Person 1	
Person 2	
Person 3	

Add this to your Climate Action Guide.

At any point on this journey, consider connecting with people and talking about climate action. Don't be disappointed if they don't all respond positively straight away. From my own personal experience, I needed to hear about meditation a few times from several friends before I tried it. When people hear about your success, they will start to believe that they can do the same things. It promotes the idea of "If they can do it, so can I." This doesn't mean "Now I've made the change, you should, too"; instead, the message is "This is my story of how I made positive change." In your story, include any difficulties, and then how you found the solution to achieve success.

Set up a Climate Circle

Another way of building community and sustaining momentum is to create a small group of people who meet on a semi-regular basis, much like a book club. These could be people from your connection list, such as friends, family, or acquaintances. I've found it's very helpful to have a few people whom I can check in with regularly. You can track your progress and help each other with positive feedback and helpful advice, discuss any challenges, provide encouragement, and celebrate achievements.

Consider meeting monthly or every second month. Life is often busy, so find the timing that is convenient with everyone. Meet in person at home, a café or library, online, or other setting. I'm on an organizing committee that involves six people from across the city and one person from another country, and our online meetings work well. Having a group of about 4-8 people is good, as this means a gathering can continue if one person can't make it. It's also small enough to give everyone the chance to contribute and feel heard.

This approach is not new; having small groups to create social change has been around for a long time, and they are being used effectively today. There is information on how to set up and run a climate circle at www.climate-action.org. The intention is to create an *enjoyable* way of being involved, build community, and keep the momentum moving forward for positive change. Some people in your group may encourage others to take action. If one of these people starts their own circle, this could lead to a chain reaction and create a cascade of new people becoming involved.

In his book *Cascades*, Greg Satell wrote that the role of a leader is now less important than the growing power of networks. If only 5,000 people encouraged three people to take action, and each of these told three people, then after ten steps, the number of people involved would be over 295 million. When we get millions of people involved in unified actions around the world, we will change the course of history.

Climate circles can contribute to unstoppable momentum.

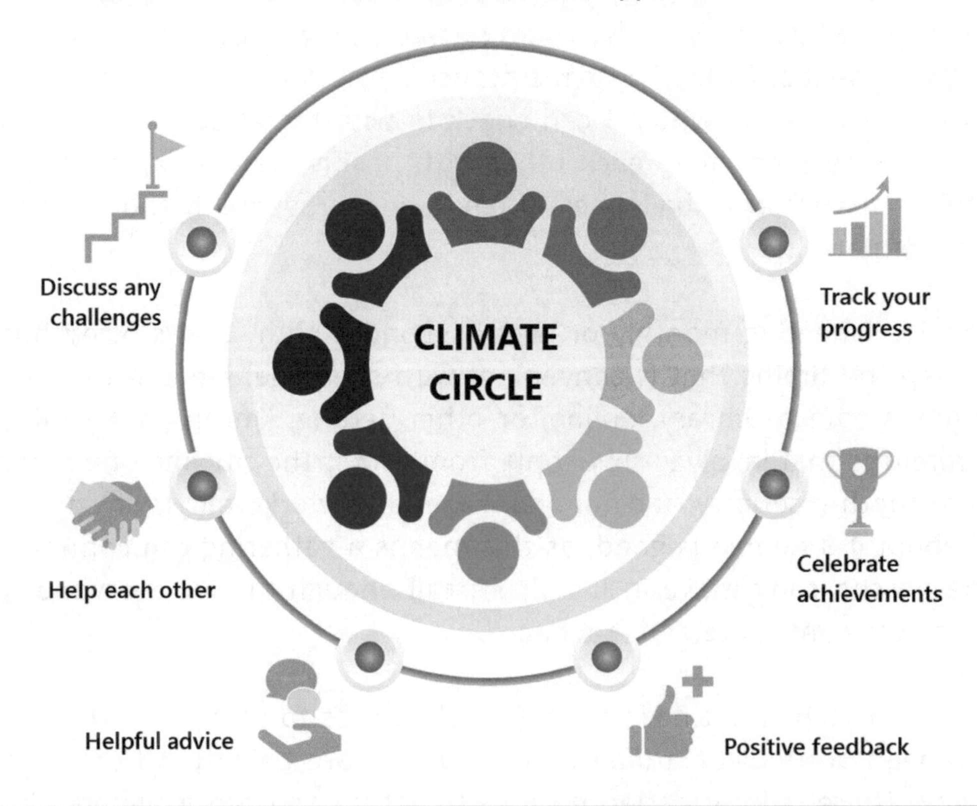

PART 3

Implementing the Actions

"There are multiple, feasible and effective options to reduce greenhouse gas emissions, and they are available now."

—Intergovernmental Panel on Climate Change (IPCC), Sixth Assessment Report (2021)

STEP 6. IMPLEMENTATION

The importance of sustainable development

"It is important to acknowledge that the climate change is part of a wider series of issues regarding the way people interact with the natural world. These issues include chemical and plastic pollution, biodiversity loss through deforestation, ecosystem destruction, and species extinction.[34] Humanity also faces people-related problems, which include poverty, hunger, access to clean water, education, health care, and energy, as well as gender and racial inequality.

Three Pillars of Sustainable Development

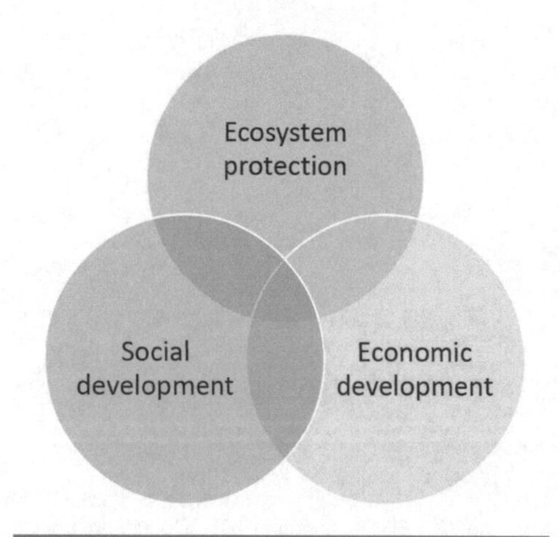

Source: United Nations 2005 World Summit Outcome

One of the first international attempts to address these issues was to convene the World Commission on Environment and Development in 1987.[35]

The report that came from this meeting, entitled *Our Common Future*, described sustainable development as the processes of economic and social development within the planet's ecological means.[36]

They concluded, "Sustainable development seeks to meet the needs and aspirations of the present without compromising the ability to meet those of the future."[37] The commission recommended the establishment of a United Nations Programme on Sustainable Development.[38] In 2015, 193 countries agreed to 17 Sustainable Development Goals (SDGs) with the aim of achieving significant progress by 2030.[39] According to the IPCC, the sustainable development goals provide a framework for climate action within the different aspects of sustainability.[40]

Seventeen Sustainable Development Goals

Source: United Nations Department of Economic and Social Affairs Sustainable Development[41]

Goal 13 is specific to climate action, while many others are interconnected to climate change. For that reason, the Sustainable Development Goals will be integrated into the objectives for each of the implementation sections: food, production and consumption, energy, and transportation. Knowing about the goals is also important because we can all play a role in improving the lives of people in our own country and all around the world. You can find more information about the goals at sdgs.un.org/goals.

Tips to Help You Be Successful

Before we start implementing the actions, I've included some tips and tricks that I have collected along the way, to make the process easier and more successful.

Be the agent of your own change

Sometimes events intervene in our lives from an external source such as economic downturn, job loss, or the death of a loved one. At other times we might move back home, change careers, or get married. These significant events can be referred to as "moments of change."[42] Climate change, while caused by human activities, is also affecting and disrupting the human way of life. To respond, we can recraft our lifestyles by altering our existing habits and behavior patterns. Change is not always easy, but it doesn't have to be as hard as we sometimes make it. Reading the book *Atomic Habits* by James Clear helped me make many small changes that gradually built up and made a big difference in my life. I'll outline some of the tips from this book and how they work.

Reward success

If you are starting an action that you think you might have a bit of difficulty completing, then another tip from *Atomic Habits* is to plan to reward yourself. This could occur when you reach a milestone or the end of an action.

Tracking progress

Tracking the progress of your actions can be simple and can help with motivation, using rewards, and making sure you succeed. For example, if you are working on reducing your energy consumption, you could measure your electricity usage at the beginning. Then decide on the different actions you will take to reduce your usage and mark them off when they are finished. In the end, you can measure your usage to see how you did.

Engineering easy wins

I put a lot of hard work into many aspects of my life. However, I also look for ways to make things easier, quicker, or more enjoyable. A great way to create momentum is to pick something easy and score a win. At the beginning of each section, I'll start with some of the easiest options to implement so you can accomplish some quick wins. You can break difficult actions into easily achievable parts and then tackle them one by one.

Be generous with time

You may have heard that it takes 21, 30, or even 45 days to create a new habit. A study on how habits are formed revealed that the average participant took 66 days before a habit became automatic.[43] Therefore, look at giving yourself enough time to monitor your progress and be understanding if it takes longer than you initially thought to modify your lifestyle.

Planning makes it easier

A little bit of planning can simplify the doing. This is especially true when you want to change an existing habit or do something completely new or complicated. For example, to change what you are eating, a bit of planning might make it easier. If you eat with other people, you can involve them in the meal choices and when the changes will take place. Then with the meal plan decided, you can find new recipes and organize shopping. At the start of each new action, consider what planning might make your efforts more effective.

Commitment

There is a speech by a motivational speaker, Inky Johnson, that I particularly like. He says, "I'm talking about the real level of commitment. Not the commitment that falls in line if everything goes right. I'm speaking of the commitment that says, 'I am going to stay true to what I said I would do, long after the mood that I've said it in has left.'" How do successful people create this level of commitment, and can it be developed? Professor of psychology Angela Duckworth and her team reviewed performance to identify what made people succeed. There was one characteristic that those left standing all had in common. It wasn't social intelligence, good looks, talent, physical ability, or a high IQ. In her book *Grit: The Power of Passion and Perseverance*, Professor Duckworth explains, "In sum, no matter the domain, the highly successful had a kind of ferocious determination that played out in two ways. First, these exemplars were unusually resilient and hardworking. Second, they knew in a very, very deep way what it was they wanted. They not only had determination, but they also had direction. It was this combination of passion and perseverance that made high achievers special. In a word, they had grit."[44]

Growth mindset

The good news is that you can enhance your grit. As Professor Carol Dweck of Stanford University explained, "New research shows that the brain is more like a muscle—it changes and gets stronger when you use it."[45] In her book *Mindset: Changing the Way You Think to Fulfil Your Potential*, Dweck explains, "When you learn new things, these tiny connections in the brain actually multiply and get stronger. The more that you challenge your mind to learn, the more your brain cells grow, with the result being a stronger, smarter brain."[46] Dweck found that by teaching children about the brain and how the ability to learn is not fixed, they were likely to be more committed and persevere through challenges. Her research showed that when groups of students who chronically underperformed were taught in a growth mindset environment, they transformed into some of the highest performing students in their region.[47]

Embracing and overcoming failure

Matthew Syed, the author of *Black Box Thinking*, wrote that learning from failure can be "the most powerful engine of progress." If you trip and fall, I encourage you to get back up, learn from your experience, and keep going.

Using motivators

Many climate actions are listed in this book; some may not work out the way you intended. If things become difficult, consider why you decided to take action on the climate crisis. Some of my reasons include (1) not to be a bystander, (2) to make the world a better place, (3) to protect people and all life from worsening climate impacts, and (4) to take responsibility to be a good caretaker of our home on Earth, passing on a sustainable way of life to the next generation. Your reasons could be more specific, perhaps about protecting your children, family, or friends. Consider writing out your motivators because they may help protect you from uncertainty or doubt.

Community

You might join a variety of organizations where you can meet other people and take part in community actions. This is also a good way to socialize, make friends, and even find a new community to be part of.

Help and include others

Once you are on your way with reducing emissions, don't forget to help others. Share your experiences and assist other people where possible.

Accountability buddy

I've found it's very helpful to have someone (or a group) that you can check in with regularly, to track the progress of your plan. This can be a partner or friend you are in regular contact with. You can help each other with feedback, advice, and encouragement.

Food

Source: luoman via iStock

ACTIONS – FOOD

Food is a critical building block for life. However, our ability to feed people is now increasingly under threat from droughts, floods, fires, and desertification—climate change makes all of these worse.[48]

Sources: Our World in Data, The Lancet, and Intergovernmental Panel on Climate Change (IPCC)[49]

Situation

Many advances in agriculture mean that it has never been easier for some people to get food. When some people are hungry, they simply go to a store and buy food items that are ready to cook. Alternatively, they can go to a restaurant and enjoy a meal not only prepared for them but brought to their table. If they don't want to go out, they can have cooked meals delivered to their doorstep.

For other people who rely on growing their own food, if the seasons are bad or they face droughts, floods, or fires, there could be little or no food at all.

Globally, a staggering one in three people (2.3 billion) are affected by moderate or severe food insecurity.[50] Of these, 870 million people face hunger, and 149 million children suffer from malnutrition.[51]

The United Nations says we can feed the world with the amount of food currently being produced.[52] How is that possible? Many factors come together in this complex issue but there are solutions we can all take part in. For example, more than one third of all food produced worldwide is lost or wasted each year, estimated at 1.3 billion metric tons and worth around US$1 trillion.[53] Food loss happens because of issues with harvesting, storage, and transport, or food is wasted by retailers, food service, or consumers.[54] Individuals and industry make many choices about food each day that contribute to food loss and waste, as well as emissions.

One way of feeding a growing population, for instance, has been to cut down more forests to expand the land used by agriculture. Humanity has already cleared an area about the size of South America for crops and an area the size of Africa for livestock.[55] This is the major cause of biodiversity loss and is one of the major drivers of climate change.[56]

It is estimated that the food production system is responsible for 25% of global greenhouse gas emissions every year.[57] With a growing population, global food production is projected to increase 50% by 2050.[58] Without making vital changes now, this will have a significant impact on nature and emissions. Instead of going down, emissions from agriculture are expected to increase by 30% by 2050 (IPCC).[59]

But it's possible that this number could go down. Remember that the relationship that humanity has to food is about decisions made. Indeed, the single most powerful thing we can do to change how food affects the climate crisis is to make different choices. The IPCC reports that if more people switch to healthy food lifestyles, it will lead to significantly reduced emissions.[60] Healthy food decisions will lead to low emissions.

We the people have the power to create and influence positive change. When individuals make different food choices, the food industry will have to adjust the supply to meet the shift in demand. Industry professionals will then make different choices, and collectively, this will reduce emissions. The following section includes actions that will help support farmers and the food industry, reduce hunger, and achieve significantly lower emissions.

We Can Feed All People If We Choose To!

Source: Tinnakorn Jorruang on iStock

Objectives

These objectives are aimed at reducing emissions related to food as well as supporting the UN's Sustainable Development Goals (SDGs):

- Encourage and support farmers and all the food supply chain to reduce their emissions by at least 45% by 2030, then to net-zero as quickly as possible.[61]
- End hunger and malnutrition (SDG 2.1, 2.2).
- Halve per capita global food loss and food waste (SDG 12.3).
- Encourage sustainable food production systems and resilient agricultural practices that increase productivity and production (SDG 2.3, 2.4).
- Indigenous lands should be protected from encroachment of agriculture and forestry in accordance with the United Nations Declaration on the Rights of Indigenous Peoples.[62]

These are some of the interconnected solutions we need to create a better future. You can find out more about the Sustainable Development Goals at https://sdgs.un.org/

How to achieve the objectives

Based on research efforts, several connected solutions are required to improve food systems:[63]

1. Consumers switch to a lower-emissions food lifestyle.
2. Farmers, transportation, and food production reduce greenhouse gas emissions and food loss.
3. Retail, food service, and consumers reduce food waste.
4. Government creates policies which support farmers.
5. We protect forests and reduce the expansion of agricultural lands.
6. We share research and methods to support agriculture improvement in production while lowering emissions.

We will explore how we can reduce emissions and food waste, as well as communicate and encourage positive change and collaboration with consumers, farmers, researchers, producers, retailers, and government. The additional aim is to give you the information you need to make educated decisions on creating a healthy, sustainable, and low-emissions food lifestyle.

Personal Actions

The decisions we make as consumers can significantly influence emissions from the food sector. If we can become more thoughtful about the food we put on our plates, the choices we make will help decide the future. The extent to which people will be able to adjust their food lifestyles will vary according to people's circumstances. The following suggested changes are intended to be applied while respecting regional contexts, including cultural and religious norms, as well as acknowledging different income levels and personal preferences.

Our Choices Make a Difference

Source: Anastasia Gubinskaya on iStock

Eating Food with a Lower Carbon Footprint

Not all food creates the same emissions

Researchers investigated the emissions from producing and consuming foods in more than 100 countries.[64] They started at the farm and measured emissions through the supply chain to the consumer's plate. They found that some foods have a bigger impact than others. The graph below shows the greenhouse gas emissions created to produce 1 kilogram (2.2 pounds) of several animal-based foods.[65] The first thing to notice is that some foods have a lot more emissions than others. To produce 1 kg of beef, emissions equal to 60 kg of CO2 are made. Emissions are higher for beef, lamb, and dairy because large amounts of methane are created by the animals' digestive process.[66]

Animal-based Food Emissions Impact of 1 kg (2.2 lb.) of Each Food Type

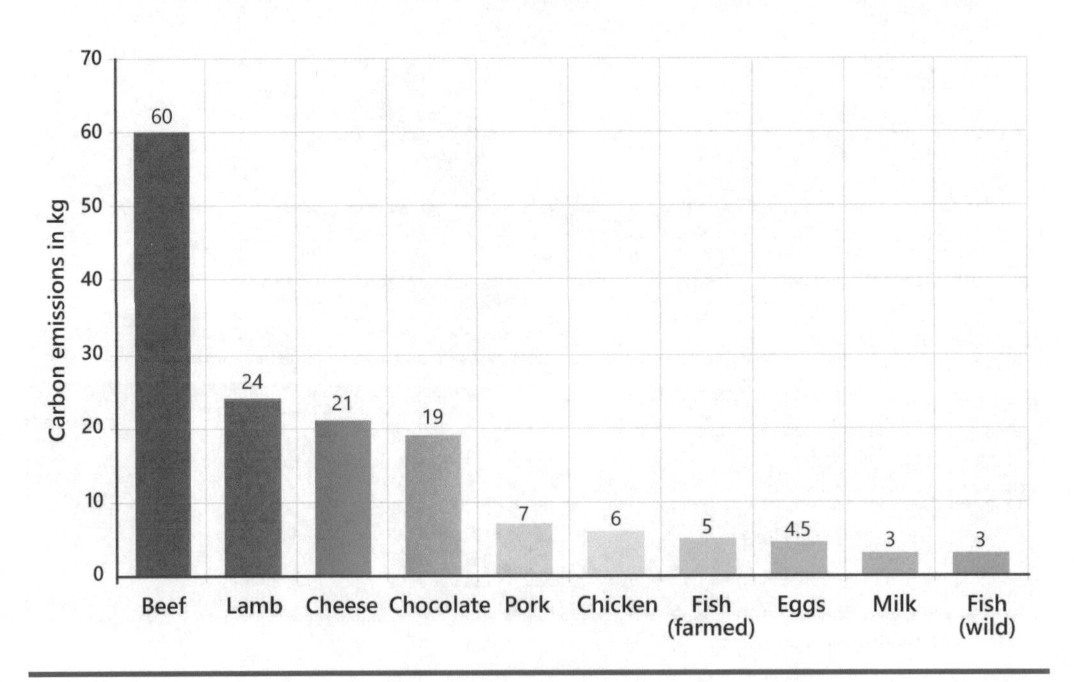

Source: Joseph Poore and Thomas Nemecek in Science, (AAAS), via Our World in Data

In 2021, the average annual beef consumption in the United States was 26 kg (58 pounds) per person.[67] If you multiply that by 60 kg of carbon, this equals approximately 1.5 metric tons of carbon emissions for each person in the United States for beef alone. While beef, lamb, and cheese have the highest emissions, plants mostly have the lowest impact.[68]

Fruit and Vegetable Emissions Impact of 1 kg (2.2 lb.) of Each Food Type

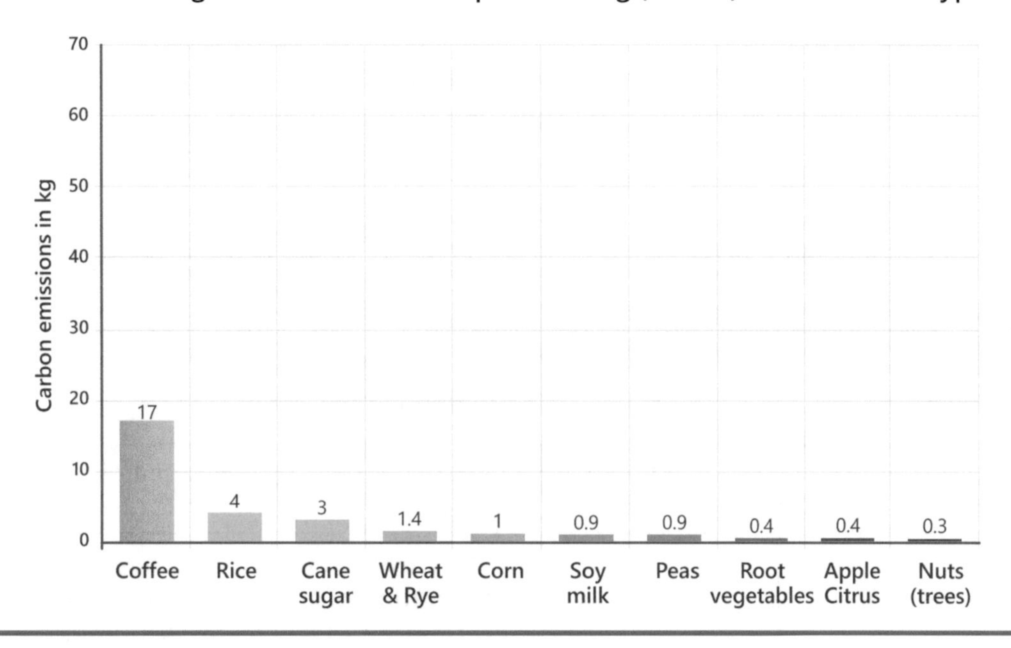

Source: Joseph Poore and Thomas Nemecek in Science (AAAS) via Our World in Data

What are the different options to reduce emissions?

Depending on the people you ask, you'll get a wide variety of answers. Note that before you change your food lifestyle, you might consider discussing it with a doctor or dietician. (Speaking of which, I am not a doctor or a dietician, and this data is intended as information and not medical advice.)

In its *Special Report on Climate Change and Land Demand Side Mitigation*, the IPCC stated with high confidence that changing food lifestyles presents "major opportunities for reducing greenhouse gas emissions from food systems and improving health outcomes."[69] The report reviewed several food lifestyles and how they compared with emissions reductions. Four of the main options are vegan, vegetarian, healthy diet, and Mediterranean.[70]

The IPCC compared what the reduction in global emissions could be if all people on the planet switched to these food lifestyles (see the following table on the next page).[71] You can choose one of these to model or inform your food lifestyle, or something else altogether.

Four Food Lifestyles and Potential Annual Emissions Reductions by 2050

FOOD LIFESTYLE	EMISSIONS REDUCTION	DESCRIPTION
Vegan	7.8 billion metric tons	Completely plant-based
Vegetarian	4.6–7.2 billion metric tons	Vegetables, fruits, grains, sugars, oils, eggs, and dairy
Healthy Diet	4.3–6.4 billion metric ton	Based on global dietary guidelines for consumption of fruits and vegetables, with reduced to low meat consumption
Mediterranean	1.2–2.3 billion metric tons	Vegetables, fruits, grains, sugars, oils, eggs, dairy, seafood, with moderate amounts of poultry, pork, lamb, and beef

Source: IPCC, Special Report on Climate Change and Land Demand Side Mitigation

We can save up to six billion metric tons of greenhouse gases each year by following healthy food guidelines. Lists of the national guidelines for many countries are available at www.fao.org/nutrition/education/food-based-dietary-guidelines. The dietary guidelines for the United States can be found at www.MyPlate.gov. This site presents information, resources, recipes, and can tailor a personalized food plan based on age, gender, height, weight, and physical activity level: www.myplate.gov/myplate-plan. Everyone is different, so consider your own circumstances and consult with a doctor, if necessary, before making changes.

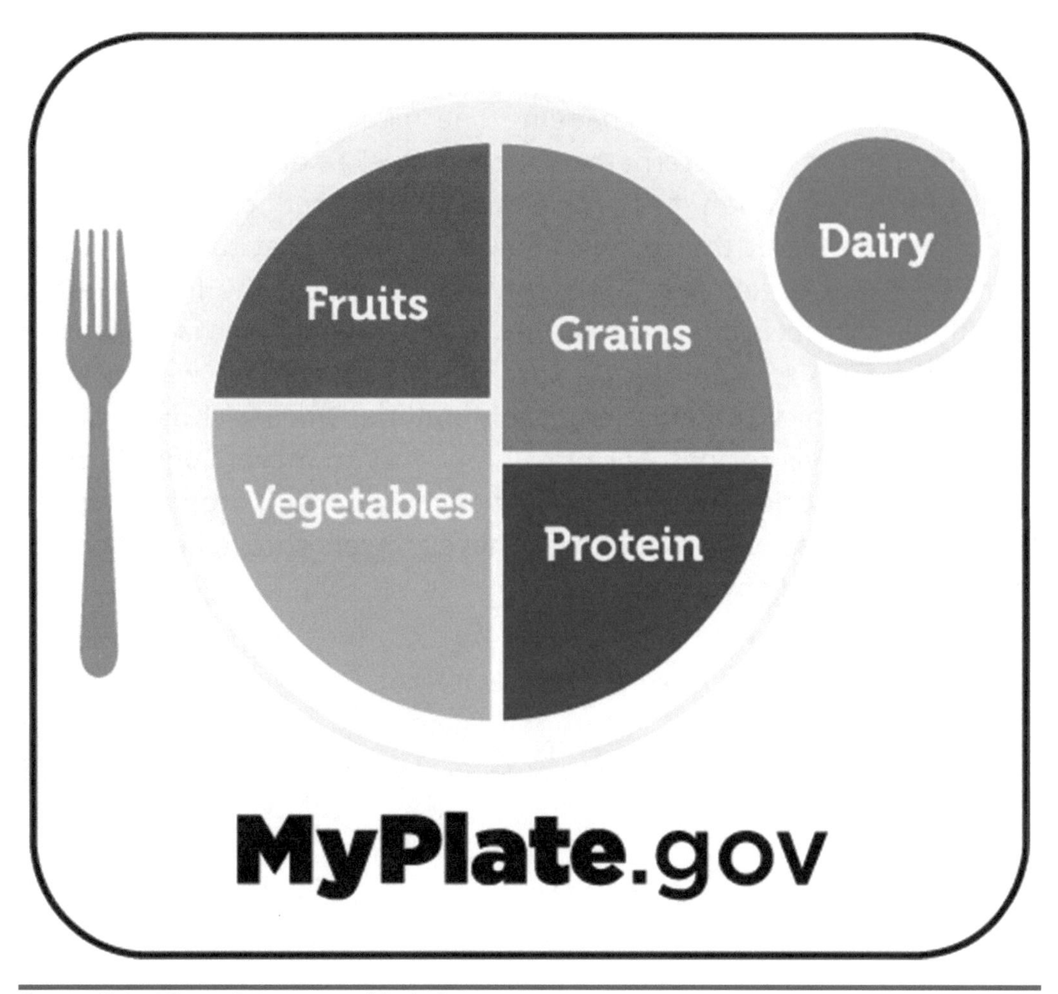

Source: United States Department of Agriculture

The guidelines note that more than 80% of people in the United States have a food lifestyle that is low in fruit and vegetables.[72] Also, many people are exceeding total protein recommendations for meats, poultry, and eggs.[73] The guidelines suggest that on balance, half of the food on our plates should be comprised of fruit and vegetables.[74] Because our choices can bring about change, healthy food decisions will lead to low emissions. The MyPlate symbol only suggests proportions of food groups, not absolute amounts, because each person has different calorie and nutritional needs.[75]

What I did

I made small adjustments, changing one meal in a week. I gradually swapped red meat for alternatives such as chicken or fish. I also started having more vegetarian meals. Over six months I reduced my red meat intake by half. I put the money I saved on less meat towards buying more fruit and vegetables. I will continue to adjust my food lifestyle to reduce my emissions even more. To make changing habits easier, review the "Tips to Help You Be Successful" section. One way to make a change is to pick the most enjoyable option. One way I did this was to make sure new recipes were tasty and easy to make. Some farmers and farming associations are already making important progress toward low- or zero-emissions produce, and these growers should be supported by consumers and governments.

Talk with people

If you share meals with other people, involve them in the meal choices, the changes to make, and when they will take place.

Substitute from high- to low-carbon foods

Little adjustments, such as eating less beef, can quickly and significantly reduce your food footprint. There are many alternatives made from vegetable protein and other products that look and taste like meat.

Reducing portion size

Reducing the amount of high-emissions food in a meal is also a step in the right direction.

Half the plate

When planning meals, aim to fill half your plate with fruit or vegetables. If those who eat more red meat than recommended by the guidelines reduce their intake, it can be healthier for them and will reduce emissions.

Planning saves time and money

With a meal plan agreed on, you can plan what you are going to eat for the week. Deciding what recipes to use will inform the ingredients you need. Making meals tasty can support change acceptance. The MyPlate website presents more meal planning resources: www.myplate.gov/resources/print-materials.

No Meat Monday – No Meat May

One way to create a new food habit is to pick one day a week and aim for low emissions. If Monday is the first day of the working week for you, then starting change might be better on a different day, like Wednesday, for example. There is also No Meat May, a challenge from a group in Australia. They have a website with lots of resources, such as recipes, meal plans, blogs, and more: www.nomeatmay.org.

Transporting Our Food

It's not only the choice of food type that is important; how it is shipped can decrease or increase emissions as well. Food can be transported in many ways. It can be produced locally, driven by truck over long distances, canned and shipped by sea, or picked fresh and air freighted. Each of these modes of transportation has a different amount of carbon emissions. Fresh food transported by air freight will likely have the highest emissions, in some cases 50 times higher than sea freight.[76]

PERSONAL ACTION — REDUCE TRANSPORTATION EMISSIONS OF FOOD

Buy in season

One of the simplest choices around reducing our emissions is buying fresh fruit and vegetables in season.

Buy local

If there is a tag or label, find out where the fruit and vegetables are coming from. By buying local or regional produce, you support farmers and businesses in your community. You also help fight emissions and pollution by reducing delivery distances for trucks and other vehicles.

What I did

I have been buying fruits and vegetables in season, which are more likely to be grown locally or from the nearby region. I have also been checking labels of fruits and vegetables to make sure they haven't been air freighted long distances.

Reducing Our Individual Food Waste

When there is food loss or waste, all the resources used to make it are squandered and the emissions were created for nothing. But those emissions still contribute to climate change.

The situation

The United Nations Environment Programme estimates that one-third of the food produced in the world for human consumption every year is lost or wasted. This amounts to 1.3 billion metric tons costing US$1 trillion.[77] In addition, estimates suggest that 8% to 10% of annual global greenhouse gas emissions are associated with food that is not consumed.[78] If food loss and waste were represented by a country, that nation would be the third largest source of emissions.

Food loss and waste

Let's consider two separate aspects to this problem:[79]

- *Food loss* occurs at the production level: at the farm, during post-harvest processing, or in the distribution stages.
- *Food waste* mainly takes place at retail and consumption stages.

Food Loss: 369 million metric tons			**Food Waste**: 931 million metric tons		
Farm	Transport	Processing Packaging	Retail	Food Service	Consumers
Not enough data available			**118 million**	**244 million**	**569 million**

As you can see from the previous table, most of the food that isn't eaten is wasted by consumers after they buy it and take it home.[80] In countries where data was available, food waste was high at the household level, estimated at 569 million metric tons per year.[81]

Setting your own objective

One of the Sustainable Development Goals set by the United Nations Programme is to halve per capita global food waste at the consumer level.[82] Consider setting an objective of reducing your food waste by 50% within the next three to six months or another suitable time frame that matches your circumstances.

Growing and composting

Think about planting some vegetables or herbs to eat. This could be in a small garden, or in pots if you are renting. Instead of throwing away your food scraps, see if you can put them in a home compost bin, or use a compost program at a community garden if there's one nearby.

What I did

We started with creating the statement "We value food and don't let it go to waste." It took about six weeks of habit formation, and it didn't always go to plan, but we forgave our mistakes and kept going. My father and I used all the ideas on the next page to reduce our food waste by 80%. While reducing emissions, we also saved money. Then we planted a small vegetable garden. It was enjoyable to plant vegetables and then eat the fresh food we had grown. You could start with one plant in one pot and see how you go.

Planning saves money

Taking a few minutes to plan what you are going to eat for the week can make the process easier. Deciding what recipes to use will help to work out the ingredients. Check what you have already, then make a shopping list. Also, avoiding impulse buys can help reduce waste.

Not typical is beautiful

We can help retailers reduce food waste by buying oddly shaped fruit and vegetables. A straight banana tastes the same as a bendy one!

Make a date

Food products often have a "sell-by" date used by the retailer which you can ignore. "Best-before" dates show when the food is at its best quality. The date you need to know is the "use-by" date, which is the last date recommended for the use of the product.[83]

Store wisely

Put new food to the rear of the shelf and push older items to the front. This applies to the refrigerator, freezer, and cupboard. Use airtight containers to keep opened food fresh in the fridge and close packets tightly.

Check freshness

A few times a week, go through the fridge and check meat, dairy, fruit, and vegetables for dates and freshness. Then prioritize eating food that could go bad soon. Use mature fruit and vegetables in smoothies or juices.

Love your leftovers

Leftovers can be kept for about three days in the refrigerator.[84] If you don't think you'll be able to eat them soon, freeze them. Leftovers can be eaten for lunch or as ingredients to make pasta sauce, stew, burritos, frittata, or soup.

Odd ingredients?

At some point you might have a few vegetables you are not sure what to do with. Ask family or friends for recipe suggestions, look online for "What can I cook with [ingredient 1] and [ingredient 2]?" or use a recipe app.

Communicate

COMMUNICATE — SHARE INFORMATION ABOUT YOUR EXPERIENCES

Conversations about food

People often want to make a positive change and reducing emissions from food is an easy way to start. You can tell people how you have changed your eating lifestyle in person or on social media. If you find a tasty low-carbon recipe, or a new low-carbon food, you can tell your friends.

Spread the word

Share what you've learned about food waste with friends, family, and colleagues. Consider writing an opinion piece for your local newspaper. Share "I Value Food" graphics and articles on social media and invite people to learn more. If you start growing vegetables or herbs, you could post about your first crop. Some people might take a photo and post on social media (consider adding hashtags: #climatechange or #lowcarbonfood).

Community groups

Do any of the organizations you are a part of serve food, such as your workplace, school, college, sporting association, or social clubs? If so, consider finding out how they manage food waste. Large organizations might have a sustainability manager. Find the person you think might be most receptive to discussing ways they could reduce waste. Using the tips in the previous section, they might be able to save food, emissions, and money.

Connect

CONNECT CONNECTING WITH FAMILY AND FRIENDS OR WITH GROUPS

Discussing food emissions and waste

Talk with your family and friends and find out what they do to reduce waste: how much are people throwing out, and how much money might they save if they throw out less?

Sharing is caring

If you have too much food that might go bad, consider sharing with family, friends, neighbors, or a charity. If you have processed or packaged food with a "use-by" label, then this is helpful when donating, so recipients know that the food is still safe to eat. If you are going away for more than a few days, check your fridge and cupboards and give away anything that might not last.

Community gardening

Are you part of a community or other social garden in your area? If you haven't gardened before, consider trying it. Growing evidence indicates that exposure to plants and green space, and particularly to gardening, is beneficial to mental and physical health.[85] Is there a vegetable garden at your school, college, or university? If so, consider taking part. You can search online for community gardens in your city. For example, community gardens in New York are listed at **www.grownyc.org/gardens/our-community-gardens**.

Volunteer with local food rescue organizations

Hundreds of organizations across the globe are actively working to rescue and redistribute safe and clean food to those in need. In some areas they are called food banks or food pantries. If you can spare the time, volunteering is a great way to impact the specific food waste challenges in your area.

CONNECT   CONNECT WITH RELIGIOUS GROUPS

Religious groups

If you belong to a specific faith community, you might see if there is a smaller group within the organization. Ask one of the organizers if they might be interested in bringing people together to promote the official position of your faith on climate change and reducing emissions.

When discussing actions that people can make, it could be easy to start with the many ways we can prevent food waste, which is aligned with the message of many faiths to end hunger and help people in need. After that, it might be helpful to identify other personal actions that people in the community would be able to easily implement. Many religions have explicitly stated their commitment for action on climate change:

- **Anglican:** acen.anglicancommunion.org/media/148818/The-World-is-our-Host-FINAL-TEXT.pdf
- **Buddhist:** fore.yale.edu/files/buddhist_climate_change_statement_5-14-15.pdf
- **Catholic:** catholicclimatecovenant.org/learn/teachings/
- **Evangelical Christian:** www.nae.net/nae-issues-call-to-action-on-creation-care/
- **Hindu:** www.hinduclimatedeclaration2015.org/english
- **Islamic:** www.islamic-relief.org/news/muslim-leaders-deliver-islamic-climate-change-declaration/
- **Jewish:** www.arcworld.org/downloads/JewishClimateCampaign%20Draft%201.pdf
- **Jain:** www.jaina.org/page/ClimateDeclaration
- **Quaker:** www.quakersintheworld.org/quakers-in-action/394/Climate-Change
- **Sikh:** www.ecosikh.org/sikh-statement-on-climate-change

Influence

Opportunities in agriculture

Food is grown on more than 600 million farms in diverse climates and ecological and cultural circumstances, on varied property sizes, and using vastly different methods.[86] So how do we find a way of reducing emissions while feeding a growing population and being sustainable? A study of 38,000 farms, as well as 1,600 processing, packaging, and retail businesses across 119 countries, looked for answers.[87] One of the key findings was that the environmental impacts from one farm can be 50 times higher than another producing the same food.[88] This indicates substantial opportunities for improvement in some areas.

Source: Markus Winkler on Unsplash

Positive change can happen by identifying and measuring not only greenhouse gas emissions, but also other ecological impacts. These include polluting runoff from the use of fertilizer and herbicides, as well as water and land management. The findings of the research "support an approach where producers monitor their own impacts, flexibly meet environmental targets by choosing from multiple practices, and communicate their impacts to consumers."[89]

Government Leadership

In each of the four Implementation sections—food, energy, transport, production, and consumption—there is an influence section which outlines ideas for basic government policy on climate action. For each of the Government Leadership sections, a generic plan outline is presented in the Appendix. If your government doesn't already have these in place, then you could try meeting with a political representative or candidate from the voting district or region you live in, as they are more likely to listen to someone who might vote for them. You could do this with family, friends, or through local political or climate action groups.

There might be organizations that support candidates whose policies include taking action on climate change in your region. An example of this is the Brand New Congress in the United States. You can get involved by signing petitions, writing letters, sending emails, protesting, and in other ways.

Your vote counts

Find out which politicians and political parties have declared that they will not take money from the fossil fuel industry. Consider voting for politicians and political parties who will act now to reduce emissions by 2030.

INFLUENCE LOBBY GOVERNMENT FOR A FOOD EMISSIONS REDUCTION PLAN

Whom do I approach?

If your government doesn't have a food emissions reduction plan already in place, then you could approach a national or state government representative. Alternatively, you could approach an official in agriculture. You could do this on your own, with friends, or through a local political or climate action group by searching online.

Make it your own

The outline to lobby government is in Appendix C: Framework for Food. You can edit this and make it your own if you like and then discuss it or send it to your local political representative. Sections are included to discuss supporting farmers to implement existing solutions, research into emissions reductions, reducing food loss, uptake of new technology, and many others.

Single issue

You could start with one issue that applies to your region or one that particularly interests you.

Look and listen

Consider doing a quick search online to find out if these ideas already exist in your state or country. Always ask questions first and actively listen. Then start a conversation about potential areas for action.

Food Manufacturing

Some food in your country might be manufactured locally, while other food might be imported. Whatever the situation, we can still influence the emissions our food makes by contacting the producer or distributor. We need to communicate with food companies and let them know what we want from our food, now and into the future. If you can get other people involved from your family, friends, school, college, workplace, and social or community group, it will have a much bigger effect. In the past, businesses have been concerned about their competition. Now is the time to make them concerned about how their own customers are demanding action on the climate crisis.

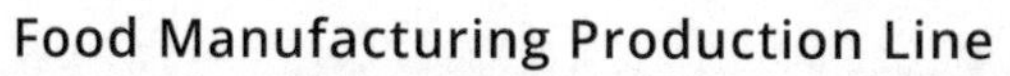

Food Manufacturing Production Line

Source: MJ_Prototype on iStock

INFLUENCE GETTING FOOD MANUFACTURERS TO REDUCE EMISSIONS

One way to make a difference could be to find out which are the most popular brands, as these are likely to have the biggest impact. You could begin by asking questions about the manufacturer's policy, objectives, and plans to reduce emissions. If they don't have any, or if they are weak or unclear, then it is reasonable as a customer to ask them to develop objectives and plans. An ongoing campaign could also involve the local media. You could ask questions using many channels:

- An email to the food producer's website
- A post, comment, or question on their social media sites
- A product review that asks for climate action now
- Posts on your own social media

Demand net-zero emissions food options

Manufacturers should cooperate with farmers, transportation companies, retailers, and the food service industry to reduce the emissions impact of food across the supply chain by 45% by 2030. They could start with one or a few products and make them low- or zero-emissions, and then gradually work through their product line as the process becomes more streamlined.

Reduce corporate emissions

Demand 45% emissions reduction on their corporate operations by 2030. Setting an objective is important, but they should also make a specific plan and put it into action now.

Better packaging

Food producers should reduce the amount of packaging used, switch to more sustainable options instead of plastic, and provide advice to customers on storing their products as well as tips to reduce food waste.

Retailers and food service

In many countries, food retailing is influenced by a few corporations. In the US, the top four retailers account for 40% of national grocery sales, while in the UK, the top seven food retailers account for 87% of the market.[90] While this type of concentration provides questionable outcomes for consumers, it can also make identifying the larger food retailers in your country a little easier. While all food retailers should act on reducing emissions, the largest often have more technical and financial capacity. Also, when an industry leader reduces their emissions, it means a big reduction, and this can set the direction for the rest of the market.

Source: Peter Bond on Unsplash

The food service industry can also play a pivotal role in driving change. Restaurants and fast food, especially the large chains, should start providing zero-emissions options and reducing food waste as much as possible. Consumers should reduce their own share (61%) of the estimated 931 million metric tons of food wasted every year. However, retailers still account for 13%, while food service also has work to do at 26%.[91]

Net-zero emissions food options

In many countries, supermarkets have considerable sales of their own brands, so they can also begin introducing zero-emissions food products. Food service companies could begin with creating one zero-emissions item on the menu and gradually increase the number, so customers can decide how their food choices make an impact.

Food waste

Both retail and food service companies can make a commitment to measure food waste and publicly report their findings, as well as set an objective of reducing food waste by 50% by 2030. They can also commit to donating 100% of edible, unsold food to a food bank or other charity.

Better information

Ask for an end to the confusion between “display until,” “best before,” and “use by” dates—there should only be one “use by” on a food product.

Food delivery services

The growing number of food delivery services should commit to a reduction of 45% in emissions for their entire operation by 2030.

Phase out unnecessary plastic

Retailers should phase out single-use plastic products and shopping bags. In addition, restaurants, and take-out food services should replace plastic with sustainable alternatives.

Schools

Many schools serve food, and this is a good opportunity to teach children to waste less food and value it more. This could be included in cooking classes, classroom curricula, educational materials and posters in cafeterias, farm experience opportunities, and school gardens with composting programs.

Compel

Workers in some professions, such as doctors or lawyers, need to be registered. People need to have a license to operate heavy machinery or transport dangerous goods. There is also a social license to operate, which is the ongoing acceptance of a company or industry's business practices by its employees, stakeholders, and the public.[92] This is often linked to an organization's environmental and social impact as well as their governance (ESG)[93] In recent years, an organization's commitment to sustainability and low greenhouse gas emissions has become increasingly important.[94]

One of the most significant things about a social license is that it can be taken away from an organization by its customers, and even by its employees. For example, if a retailer refuses to reduce food waste or sell low-carbon food options, then the public can stop buying from them and even protest in various ways against the organization. Our networked world is expanding the ability of consumers to make decisions and share their experiences online about products and services. We have control over the products we buy, and this gives us power. We can shift our buying patterns, communicate information, connect with each other, and demand positive action. When we combine our individual choices with other people and demand climate action, we have the power to change the way corporations and entire industries operate.

COMPEL | HELP THOSE RELUCTANT TO CHANGE

You decide what to buy

If food brands or retailers refuse to reduce waste or emissions, then you can choose not to buy their products or shop at their stores. Importantly, let them know you will no longer be buying from them because of their failure to take action on climate change. This can be done in person at a store or via email or social media. It works more effectively if you can connect with others to make a coordinated response.

Food Conclusion

This is a pivotal moment in history, and now is the time to find a balance between the way we are producing and using food, and how we are sustaining the vital ecosystems for future generations.

We can take action today because many of the solutions that we need are available right now. We can all play a part in reducing the food currently lost or wasted globally. This can contribute to improving access and affordability of food for the people who need it most.

We have discussed many ways to reduce our own greenhouse gas emissions, and some of the most effective ways are choices we make about the types of food we eat. We have also identified a wide range of options to communicate, connect with other people, influence, and compel action. Each time we stop food from being lost or wasted, we are reducing the impacts on land and water and saving energy, labor, and transportation resources, while reducing greenhouse gas emissions.

Think about the example we give to our friends, family, or children. We can promote the value of food and respect it by not wasting it. Can we make wasting food as unacceptable as littering? I believe we can.

Think about the emissions reductions you have made and how much food waste you've prevented. Remember to celebrate the progress you are making!

Production and Consumption

Source: Billy Clouse on Unsplash

ACTIONS – PRODUCTION AND CONSUMPTION

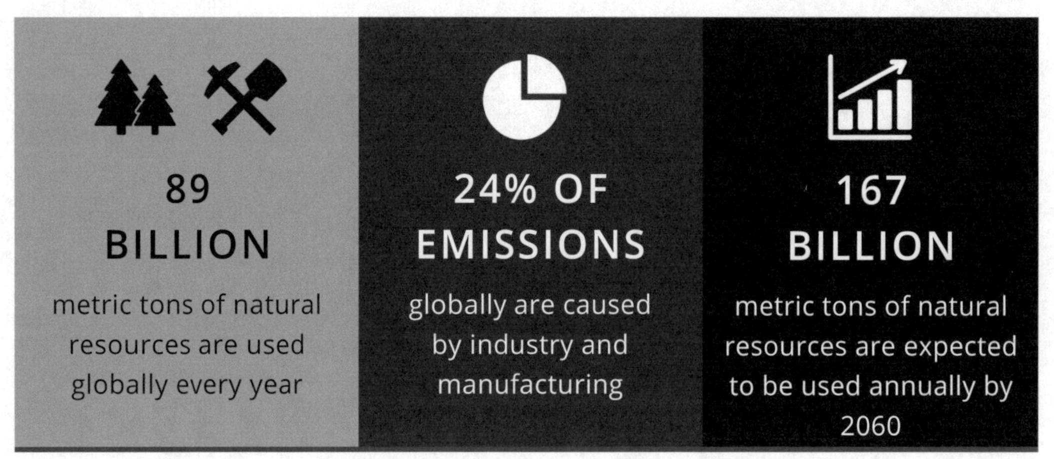

Sources: OECD and Our World in Data[95]

Since 1950, many aspects of human civilization, such as population, the economy, demand for energy, and pollution, have increased rapidly.[96] Global demand on natural resources also became higher than what ecosystems could regenerate in one year.[97] This impact of people on the earth is referred to as the ecological footprint.[98] It can be calculated by adding up total demand for crops, fish, livestock, fiber, timber, land used for habitation and roads, and waste such as carbon dioxide.[99]

In the same way that a person can go into debt with a credit card, humanity is going into debt by using more than the earth can provide—to be precise, we use an amount of ecological resources equivalent to 1.7 Earths each year.[100] This is degrading ecosystems and causing major impacts on people.

Our Ecological Footprint

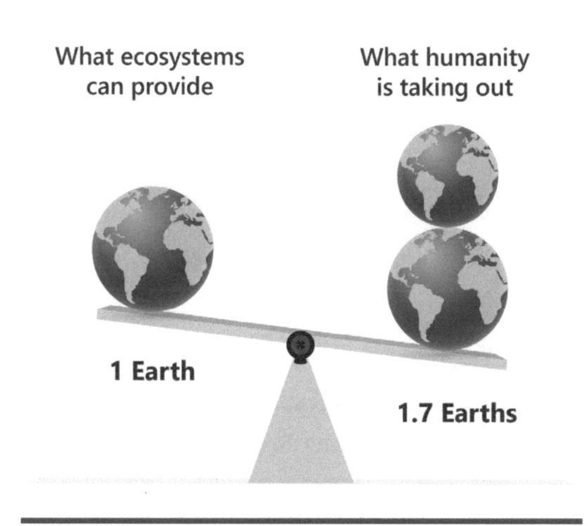

How Many Earths We Would Need

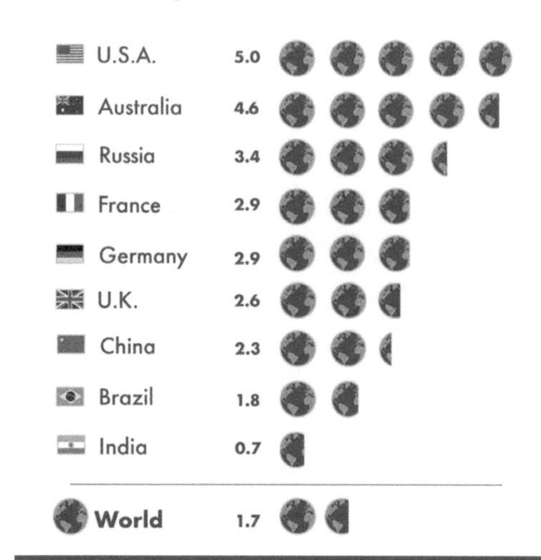

Source: Global Footprint Network 2021

This way of life is illogical and unsustainable. By looking at the ecological deficit for each country, it shows how many Earths we would need if everyone lived the same lifestyle. We can see that the high-income countries have the highest rates. For example, if everyone lived like people in Australia or America, we would need more than four Earths to provide all the ecological resources needed.

Situation

We use more than 89 billion metric tons of natural resources annually—this includes agriculture, forestry, metals, minerals, and fossil fuels.[101] Turning these into materials and products causes 13 billion metric tons of greenhouse gas emissions every year.[102] This is caused by a wasteful linear, or straight-line, economic model as illustrated on the next page.[103] It operates by taking natural resources to make products, which are used and then thrown away at some point in a process described as "take, make, waste," as illustrated below.[104] Globally, only 13% of products are recycled each year.[105]

Linear Economy (Take, Make, Waste)

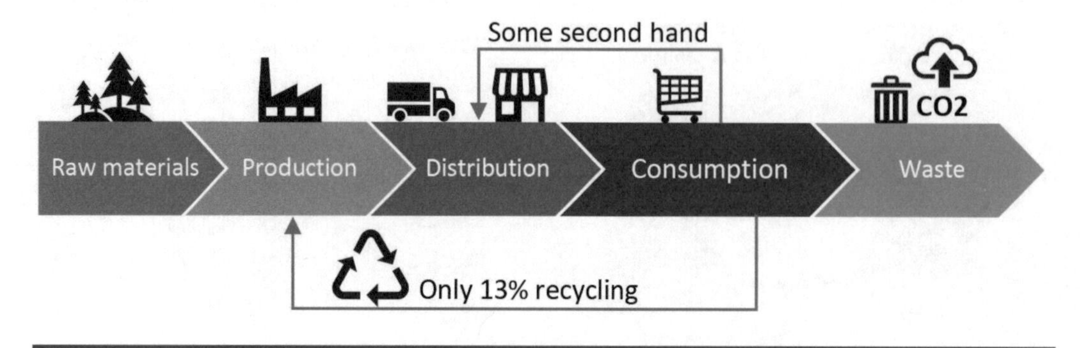

Source: Based on United Nations Industrial Development Organization, Driving Towards Circularity

The climate crisis has exposed that our current way of production and consumption is outdated and unsuitable for our modern way of life. With the global population expected to increase to 9.6 billion by 2050, the extraction and use of natural resources is set to nearly double.[106] This is an urgent problem that needs to be fixed now.

Some industries are only just developing the new technology needed, while others are doing little. Ignoring the problem is not a solution, and "business as usual" must end. We need to extract fewer materials and use what we already have. To stop climate change and the destruction of vital ecosystems, we need to make production and consumption sustainable.

The "take, make, waste" model should be replaced with a circular economy. In this new model, products are designed for durability and made to be taken apart, which makes them easier to repair.[107] Products are reused, repaired, refurbished, or taken apart and recycled. This means fewer natural resources are extracted because materials are used around and around the circular economy, as illustrated below.[108]

Simplified Sustainable Circular Economy

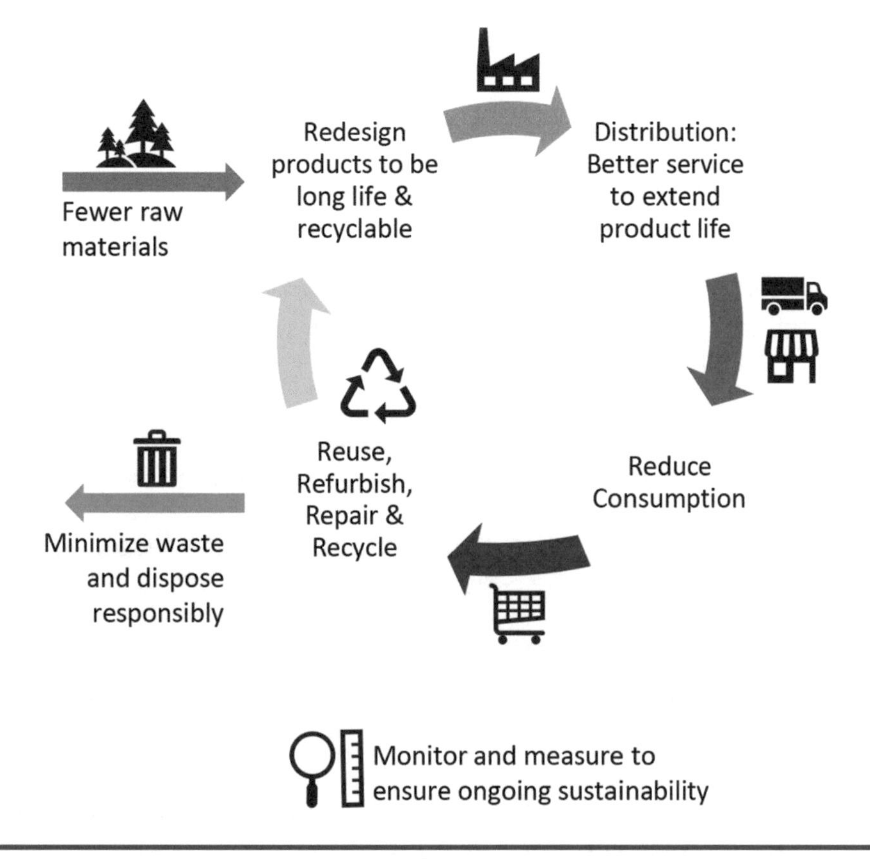

Source: Based on United Nations Industrial Development Organization, Circular Economy

The outcome is fewer greenhouse gas emissions and less impact on ecosystems and biodiversity. This more closely matches how natural systems efficiently use energy. The more we reduce our use of materials, the more sustainable our consumption will be.

Objectives

These objectives are aimed at reducing emissions as well as supporting the UN Sustainable Development Goals that relate to production and consumption:

- Encourage and support industry and manufacturing to reduce emissions by 45% by 2030, then to net-zero as quickly as possible.[109] This also applies to retailers, service providers, and all businesses.
- Implement a framework of sustainable consumption and production, with developed countries taking the lead (SDG 12.1).
- Reduce waste and sustainably manage natural resources (SDG 12.2, 12.5).
- Improve global resource efficiency in consumption and production, and decouple economic growth from environmental degradation (SDG 8.4).
- Ensure people have information for sustainable lifestyles and climate change action (SDG 12.8, 13.3).
- Support developing countries in strengthening their scientific and technological capacity to move toward sustainability (SDG 12a).

Some of the interconnected solutions we need to create a better future and the Sustainable Development Goals are available at **https://sdgs.un.org/**

How to achieve the objectives

Several connected solutions are required for production and consumption:

1. Consumers align to a low-emissions lifestyle.
2. Governments show leadership by developing policies and a plan for industry to reduce emissions and to establish a circular economy.
3. Heavy industry, manufacturers, retailers, businesses, and service providers establish emissions reduction objectives and plans, as well as apply existing renewable energy solutions.
4. Industry and government collaborate with researchers to develop new technical solutions to lower emissions.
5. Government and industry provide consumers with accurate information to participate in the circular economy with product choices.
6. Solutions are shared with developing countries, with training and funding.

Since 2009, several countries and the European Union have decided to begin implementing a circular economy.[110] Now is the time to end our throwaway culture.

Personal Actions

Product marketing often tells us we must have more, and that the newest shiny product is the key to becoming noticed and popular. The thing about a lie is that when we stop believing them, then the fog lifts and we can clearly see the real world around us. The head of sustainability at IKEA noted, "In the west, we have probably hit peak stuff. We talk about peak oil. I'd say we've hit peak red meat, peak sugar, peak stuff."[111]

A report titled "The Growing Power of Consumers" noted that "consumers' power is consolidating with improving access to information, an ever-widening choice of goods and services, and opportunities to share their experiences more widely."[112] The report also noted that consumers are "expecting to be given the opportunity to shape the products and services they consume. Consumers have been given a voice and they expect it to be heard."[113] Corporations and industries will take notice if we stop buying their products and services, because their profits will fall. We can reduce corporate profits with our purchase decisions until corporations reduce their emissions. This is how we can achieve system change and solve the climate crisis.

Support a sustainable circular economy

The amount each one of us consumes matters more now than at any time in history. This is because when we spend money on products, it has an additional cost in terms of natural resource extraction and greenhouse gas emissions. People, communities, business, industry, and governments at all levels shape the future every time they spend money.

We Can Influence the Future

Source: Joshua Rawson-Harris on Unsplash

PERSONAL ACTION SUPPORT A SUSTAINABLE CIRCULAR ECONOMY

Reduce

The first step is to reject unnecessary consumerism and the wasteful linear economy. Our choices make a difference in making a better future. Consider what you purchase and consume what you need. We can still have fun, but many people in high-income countries could consider purchasing less. To find the ecological footprint of your country, open the following link. If the ecological footprint of your country is greater than "1," consider reducing your consumption unless you are already purchasing at a low level:
www.overshootday.org/how-many-earths-or-countries-do-we-need/

Refuse

Take your own reusable shopping bag and refuse plastic bags. Stop using single use plastics such as straws, plates, containers, and cutlery. Demand renewable options from shops and take-out food outlets. A list of alternatives is presented in the following table in the "What I did" section.

Reuse

Buying products second-hand gives them a new life. Sell or give away your unwanted items to family, friends, or charity instead of throwing them out.

Repair and refurbish

Many electrical items can be repaired, clothes can be mended, and furniture can be reupholstered. If you are not sure, ask family or a friend, or search online.

Recycle

Recycling helps reuse the materials we already have, which means fewer natural resources, less rubbish in landfills, rivers, and oceans. Buy products which have a recycle symbol for their packaging. Ensure that you recycle every item you can. Ask your local government or recycling center about which items they will take.

Plastic problems

Global emissions from the plastic lifecycle in 2015 were 1.7 billion metric tons of CO2 equivalent. This is expected to grow to 6.5 billion metric tons annually by 2050 under business as usual.[114] Plastic products can last between ten and 1,000 years. This is useful for long-term products used in buildings, but for packaging that is only used once and then thrown away, it contributes to a massive accumulation of plastic.[115] Combine this with the fact that globally, 90% of all plastic made each year is not recycled, and you can understand why landfills and oceans are drowning in plastic.[116] As you can see in the chart below, packaging is the biggest generator of plastic and represents a great opportunity for positive change.[117] We can all choose to use less plastic.

Plastic Waste Generation by Industrial Sector (2015) in Millions of Tons

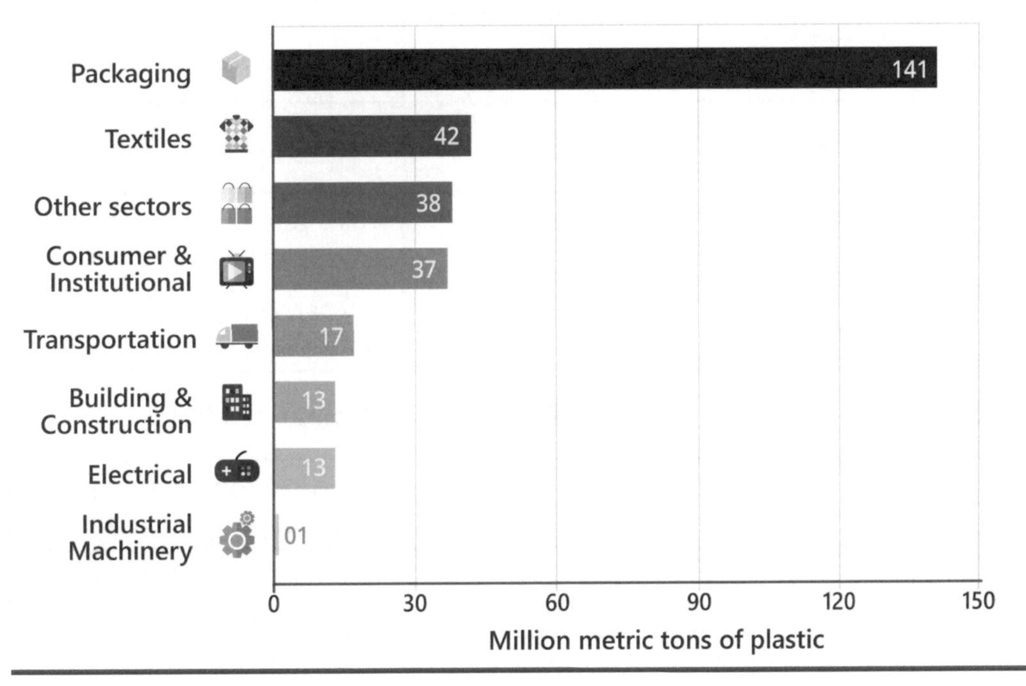

Source: R. Geyer, J. R Jambeck and K. L. Law (2017); chart based on Our World in Data

What I did

I take my own bags shopping and have stopped using plastic bags. I have a reusable coffee cup and use my own water bottle. I say "no" to plastic utensils. I traded my plastic toothbrush for bamboo (you need to wipe them dry after use). And I have cut out many other single-use plastics. If we choose to use less plastic, then we can significantly reduce emissions and pollution. Below are some common products that use plastic along with their alternatives.

Common Plastic Products, Decomposition Times, and Alternatives

ITEM	DESCRIPTION	DECOMPOSITION	ALTERNATIVES
	Plastic bag	10–20 years	Take your own bags shopping
	Polyester wet wipes	50–100 years	Cotton or hemp wet wipes
	Styrofoam	50–100 years	Use a reusable coffee cup
	Plastic straw	50–450 years	Take your own metal straw, or don't use one
	Plastic bottles	50–450 years	Try carrying a reusable bottle
	Plastic toothbrush	450–500 years	Bamboo toothbrush
	Plastic diaper	450–600 years	Consider compostable diapers
	Plastic utensils	1,000 years	Use your own utensils or ask for renewable options

Source: US National Park Service, National Oceanic and Atmospheric Administration (NOAA) and National Geographic

How to reduce emissions from consumption

PERSONAL ACTION MAKE LOW-EMISSIONS SUSTAINABLE CHOICES

Demand renewable options to replace plastic

Buy and demand products with sustainable non-plastic packaging. Several types of renewable packaging are being made. One option is packaging made from waste sugar cane pulp and bamboo.[118]

Switch away from high emissions and unsustainable products

Find out if paper and wood products such as furniture or building materials are sustainably sourced. Or do they contribute to deforestation and greenhouse gas emissions? Common items such as toilet paper, tissues, and paper can all be made from recycled materials or from materials harvested from sustainable plantations. Look for symbols to show 100% recycled or plantation-sourced products. You can search online for "Eco-friendly [product type] in [country/region]," for example, "Eco-friendly toilet paper in America."

Net-zero carbon brands and companies

Look for brands and organizations that are making changes now to reduce their emissions and switch to their products and services if suitable. Watch for companies that are promising but not changing right now.

Building or renovating

If you are going to build or renovate a home, ask the architect or builder if they can source low-emissions materials and forest-friendly timber.

Net-zero carbon events

If you're organizing a birthday party or wedding, think about ways to reduce emissions. This could include food choices, non-plastic items, and energy use.

Communicate

COMMUNICATE SHARE INFORMATION ABOUT YOUR EXPERIENCES

Contact your brands

Many brands have a presence on social media. You can go online and ask them about reducing emissions; let them know you think it's important and they should do something about it. This could be for clothes, shoes, books, appliances, electronics, hardware, housewares, furniture, or office supplies, and in online stores. It's good to be polite, but you can also be firm and let them know you are prepared to buy a different brand if they are unwilling to make positive change.

Don't forget services

Tell your service providers that you are interested in emissions reductions and ask what they are planning to do. This could apply to your streaming service, internet provider, gym, pet grooming, mechanic, sporting club, or doctor. It can also apply to schools, the post office, and government agencies.

Ask them about it

When you go into a store, you can ask them what progress they are making in reducing emissions and stocking low-emissions and forest-friendly products.

Connect

Not only are connections between people important, but also connections within organizations big and small. Businesses must play a role in reducing emissions and reducing waste. We can transform the economy so that it works for industry and people while caring for nature.

CONNECT BUSINESSES LARGE AND SMALL

Small to medium-sized businesses

Around the world, small to medium enterprises (SMEs) represent more than 90% of businesses, 60% of employment, and up to 55% of production in developed economies.[119] Do you own or work in a small to medium-sized business? If so, consider ways you can discuss reducing carbon emissions, energy usage, food waste, and recycling, as well as discussing purchasing decisions such as using less paper or single-use plastics. In many cases, reducing emissions and waste can save money.

Large companies

Sustainability or emissions reduction initiatives or groups might already exist within your organization, or you might start one with other like-minded people. An increasing number of large companies are committing to emissions reductions aligned with a 1.5°C pathway.[120] The international organization Science Based Targets initiative (SBTi) promotes best practices in emissions reductions. They have many resources that can help you get started: **https://sciencebasedtargets.org/resources/**.

Influence

Globally, heavy industry and manufacturing create more than 13 billion metric tons of greenhouse gas emissions every year.[121] This is because coal and gas (methane) are often used to run boilers, smelters, and other heating equipment to make a range of materials which include plastic and rubber, wood and paper products, glass, cement, and refining metals such as aluminum, iron, or steel. These materials are used in construction, machinery, transportation, and infrastructure.

Government leadership

INFLUENCE — LOBBY GOVERNMENT FOR AN INDUSTRY EMISSIONS REDUCTION PLAN

Is there a plan?

You could research online to find out if your state or national government has a plan that supports industry and manufacturing to reduce their emissions.

Make it your own

The outline to lobby government is presented in Appendix D: Framework for Industry. You can edit the framework and discuss it with your local political representative or send it to them. Sections are devoted to encouraging industry to replace polluting equipment with new zero-emissions options, support emerging solutions such as using hydrogen to make steel, promoting research and development as well as information sharing.

Lobby the government

Find out who is the state or national secretary for commerce or minister for industry (leaders might have different titles depending on where you live). You might meet with a state or national politician or contact them by phone, email, or social media. You could do this on your own, with friends, or find a local political or climate action group by searching online.

Establish a circular economy

The circular economy is a new way of creating value and prosperity for consumers, retail, manufacturers, and industry. Establishing a circular economy is also an important step in reducing greenhouse gas emissions. For this to happen, products must be designed for durability, reusability, and recyclability.[122]

Products designed to fail

One of the many problems with the "take, make, waste" linear economy is that some products are deliberately designed with a short life span. This is called planned obsolescence and its brightest example is the humble light bulb. In 1924, the leading light bulb manufacturers from Britain, France, Germany, Hungry, Japan, the Netherlands, and the United States formed the Phoebus Cartel.[123] These manufacturers decided in secret to reduce the life span of the old incandescent globes from 2,500 hours to only 1,000.[124]

Burnt out light bulb

Source: Yommy8008 on iStock

The cartel developed a bulb that would reliably fail after 1,000 hours, controlled production quotas, and fixed prices despite their costs falling.[125] The Phoebus Cartel no longer exists, but their legacy lives on.

Today, manufacturers use planned obsolescence differently by making products that are difficult or impossible to repair. In a report to the United States Congress, the Federal Trade Commission concluded that consumers whose products break have limited choices. They found that many products have become harder to fix and maintain, with repairs often requiring specialized tools or difficult-to-obtain parts.[126] In Britain, the government is introducing new regulations that require making parts for electric appliances readily available, which is expected to extend the life span of some products by up to ten years.[127] Governments across the world should incentivize industries to do the right thing by consumers and future generations.

INFLUENCE LOBBY GOVERNMENT TO ESTABLISH A CIRCULAR ECONOMY

Is there a plan?

You can find out if your national government has a plan to establish a circular, low-emissions economy.

Lobby the government

On the next page are elements of a plan to establish the foundation for a circular economy. You could meet with a state or national politician or contact them by phone, email, or social media. You can download the infographic on the next page from **climate-action.org**.

5 STEPS

ESTABLISH A CIRCULAR ECONOMY

The "take, make, waste" model should be replaced with a circular economy. Products should be designed for durability, made to be taken apart, and made of recyclable materials. The result is fewer natural resources extracted because materials are used again and again in the circular economy.

1 Phase-out

Phase out plastic bags for groceries and single-use plastics and replace them with alternatives. This has already happened in many cities and countries.

2 Design for life

Manufacturers must design products to be more durable. All products from pens to cars should be designed so that they can be easily disassembled and the parts either reused, repurposed, or recycled.

3 Repair

Manufacturers must communicate the life span of products, make parts available, and extend product use through repair or refurbishment.

4 Recycle

Establish and expand recycling and sorting facilities. Industry must create objectives for minimum levels for recycling and the use of recycled materials in new products.

5 Support innovation

Governments should provide incentives for research and for implementing circular redesigned products. This new technology should be shared with developing countries.

Carbon labeling

Each time a consumer, business, or government department makes a purchasing decision, they play an important role in the economy. This is because, added up, these purchasing decisions can change the direction of a brand, organization, or even an entire industry. However, buying decisions need to be based on accurate information. For example, in many countries there is an energy rating label on electrical appliances that explains how energy efficient they are.

We should also have carbon labels that show the amount of greenhouse gas emissions created to make each product. Then a consumer could compare and determine which products have the highest and lowest emissions and then decide what to buy. Research shows that two thirds of people across Europe, the UK, and the US support carbon labeling.[128]

INFLUENCE LOBBY FOR CARBON FOOTPRINT LABELING

Ask retailers and manufacturers

Some food service and food product producers are beginning to publish the carbon footprint of their products.[129] Carbon labeling is becoming established, but it needs your support. Ask retailers and manufactures to begin carbon labeling. Consider buying products with carbon labels when possible.

Lobby your state or national government

Is there a carbon labeling program in your country that is based on a national standard? You can ask your local or national representative or do an online search for "carbon footprint labeling standard in [your country]." On the following page are the foundations of a plan to establish carbon footprint labeling. You can download the infographic on the next page **climate-action.org**

5 STEPS

SET-UP CARBON LABELLING

As consumers, we need accurate information to reduce our emissions. Each product needs to have a carbon label that shows how much greenhouse gas was emitted to make it. That way, we can compare products and choose those that have the lowest impact on the climate crisis.

1 Policy

Government should set up a national standard for carbon labeling for all products, including a streamlined process to make certification by manufacturers easy.

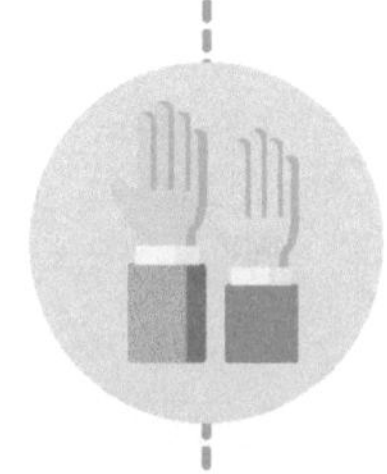

2 Voluntary

Initially, product labelling should be voluntary. The brands and companies that participate will likely be rewarded with increased sales.

3 Implementation

In the beginning, several products should be tested for each industry. Then adjustments to the certification process can be made. Gradual success will lead to widespread adoption.

4 Public education

Carbon footprint labeling should be promoted to the wider community and people encouraged to make informed purchasing decisions.

5 Help others

After high-income countries have implemented carbon labeling, they should help support developing countries to create their own programs.

Compel

In 2023, the International Energy Agency published a report on how the oil and gas industry was making transitions into renewable energy. They concluded, "Oil and gas producers account for only 1% of total clean energy investment globally."[130]

I'll be completely honest: despite what the oil and gas companies have done to prevent action on climate change, it's not my preferred option to see their shares devalued, or their assets stranded and become worthless. I would prefer that they transition their organizations to renewable energy or other areas. But they haven't, so it's up to us to encourage and compel them to change.

Make your money speak for your integrity

Consumption is not only about buying products like electronic goods or food. In fact, in the US and many other high-income countries, services make up to 70% of total domestic production.[131] We've already talked about several services, but now it's time to have a good look at some of the financial services that we use regularly.

Reporting shows that since 2015, prominent banks lent almost US$4 trillion to fossil-fuel corporations, and this money will contribute to pushing temperatures above 1.5°C.[132] We don't want banks investing our money in corporations that are contributing to the climate crisis by expanding the extraction of coal, oil, and gas.

In 2021, the United Nations convened the Net-Zero Banking Alliance (NZBA). The aim is to bring the banking industry together to "decarbonize their lending and investment portfolios on a 1.5°C climate trajectory to achieve net zero emissions by 2050."[133] While this aim is helpful, we can encourage the banking industry to stop lending and investing in the fossil fuel industry right now.

COMPEL MAKE YOUR MONEY SPEAK FOR YOUR INTEGRITY

Review your savings, credit cards, and loans

You can be an active part of the solution and prevent banks from using your money to fund fossil fuel projects by switching to an ethical bank. If you want to change banks, you might find it easier to start with something simple like a savings account or credit card. Some suggestions to make the process easier are on the following pages. Be sure to let your old bank know you are moving because they are funding fossil fuels.

This is not financial advice, so if you are unsure, please speak with a registered financial advisor.

Basic considerations for moving your money

The first step is to confirm if your bank is funding fossil fuels. You could contact them and ask if they are financing any coal, oil, or gas projects and what their plans are for the future. If they are funding fossil fuels, you could let them know that this is unacceptable and that because of this you will consider finding another bank. Find out from your existing bank if costs are involved in switching any accounts or credit cards. In addition, consideration is needed to ensure you are not financially disadvantaged by any action you may take.

If your bank makes it hard for you to find out if they are funding fossil fuels, you can search online for "Is [name of bank] financing fossil fuels?" Alternatively, your bank might be listed in the report titled "Banking on Climate Chaos" at **www.bankingonclimatechaos.org**. I looked online for my bank and found their position statement on fossil fuels, but it was unclear, so I called them. The person I spoke to didn't know, but they gave me an email address to enquire, and I received a statement that the bank does not invest in fossil fuels.

However, they had no plan to reduce their own emissions soon. I found a bank that has a commitment to have net-zero emissions by 2035. I have opened an account with them, and I am in the process of switching my banking services.

Some options to find a new bank

Listed below are a few options to find a bank that isn't funding fossil fuels:

- You can find a new bank by doing a search online using the phrase "List of sustainable banks in [your country name]."
- You could look up Certified B Corporations which have been verified based on how they create value for their employees, the local community, and the environment.[134] You can find Certified B Corporations by using their search tool at **www.bcorporation.net/en-us/find-a-b-corp/**. Type "bank" in the search field, then select your country and click "search."

Some banks promise to achieve net-zero by 2050, but this is too late. Reject any objectives that are vague—the commitment needs to be specific, include all fossil fuels (coal, oil, and gas), and have a time frame not later than 2030.

Production and Consumption Conclusion

Consumption is one of the biggest drivers of climate change, but it also impacts species extinction, water depletion, toxic pollution, and deforestation. People, businesses, and government have an opportunity to counteract this by reducing their emissions and embracing a circular economy. Each choice for a reduced emissions product or service, reusing, repairing, or recycling is a vote for moving towards a sustainable and circular economy. I hope you've considered the way you purchase products and services and how much power you have as a consumer. Remember to celebrate the progress you are making.

It's Time to Question Unrestrained Consumerism

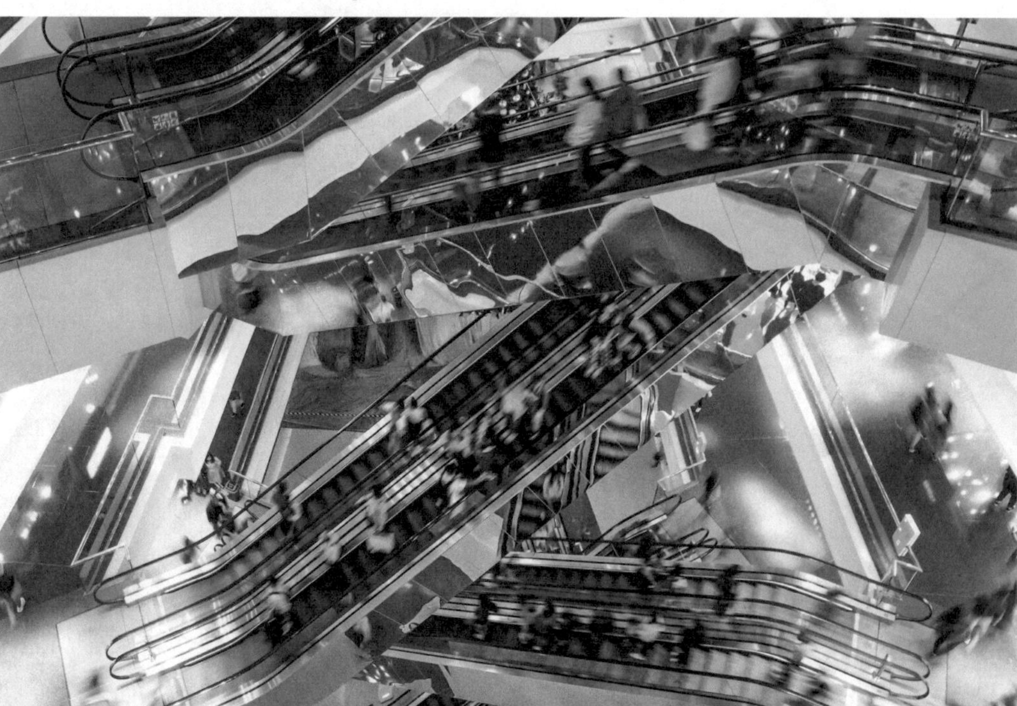

Source: Panksvatouny on iStock

Energy

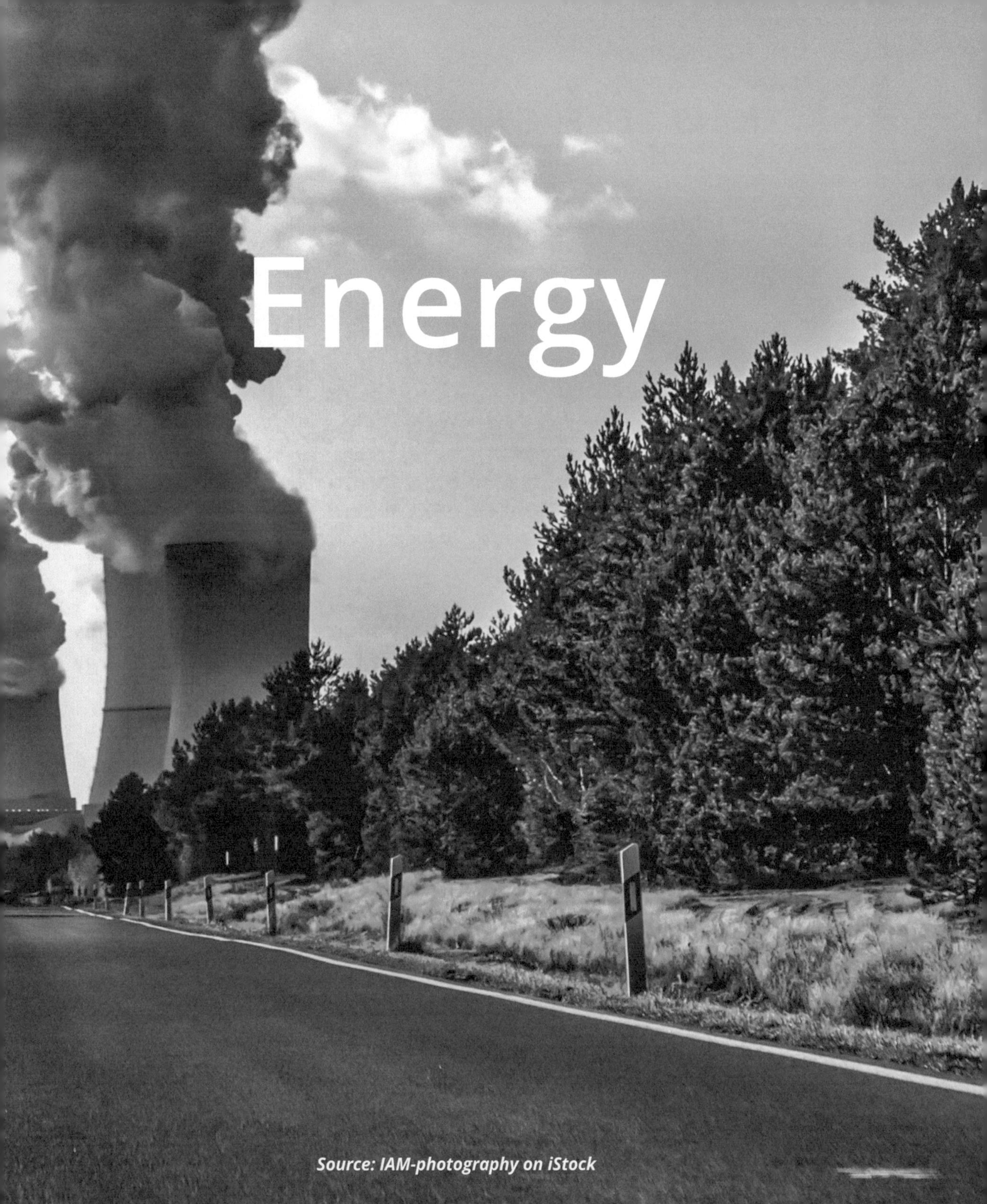

Source: IAM-photography on iStock

ACTIONS – ENERGY

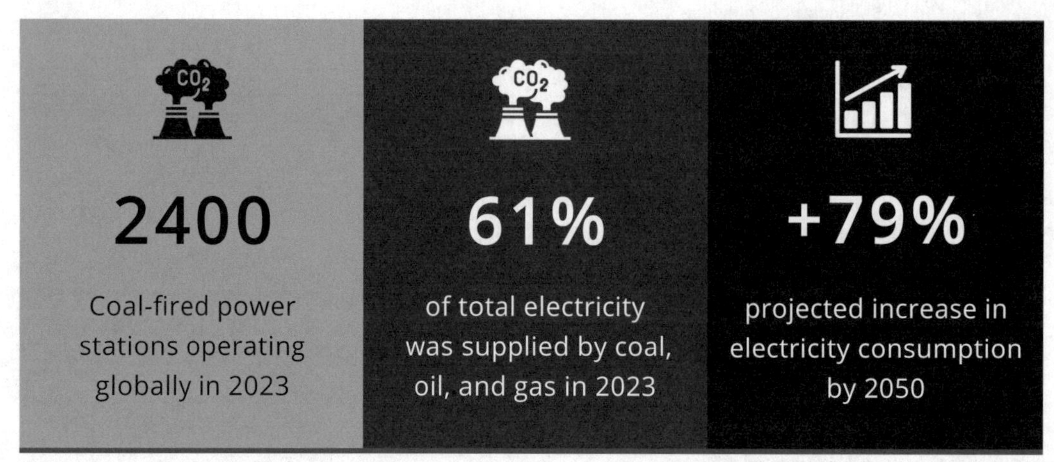

Sources: Global Energy Monitor, Our World In Data, US Energy Information Administration (EIA)[135]

Because the energy sector is the primary source of emissions, it holds the key to the climate challenge, according to the International Energy Agency (IEA).[136]

Situation

Because 789 million people have no access to electricity, many rely on burning wood for heat and cooking.[137] This causes pollution in the home, increasing risks to the health of families. For other people, access to electricity is as simple as turning on a switch. For more than 100 years, electricity has enabled lighting, heating, cooling, cooking, and has provided energy to power a wide range of appliances. Most of this electricity across the globe has been and is still produced by fossil fuels (61%), as seen in the graph on the next page.[138] In 2022, the global annual carbon emissions from electricity surged to a new all-time high.[139]

Global Electricity Generation in 2023

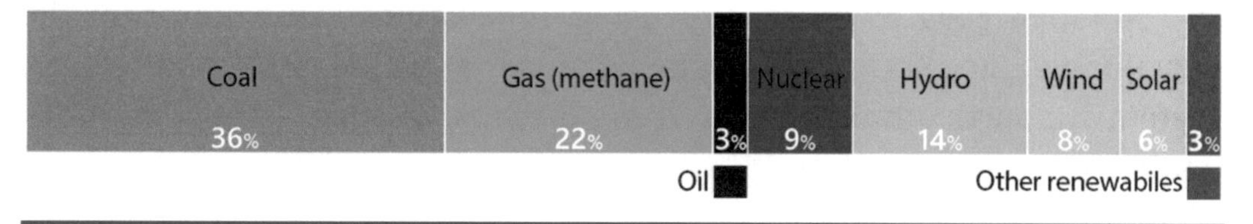

Source: Our World in Data

In the 1990s, the international community agreed to reduce emissions. But instead of investing in renewables, many fossil fuel corporations continued to expand polluting energy, so now more than 2,400 coal-fired power stations operate around the world.[140] Coal-fired power stations use old technology from the 1880s, like large steam engines.[141] Coal is burned to boil water, which turns into steam, which spins a turbine, with a generator creating electricity.[142]

Even with costs plummeting, the rollout of solar has been very slow, making up only 6% of global electricity production in 2023.[143] One of the reasons is related to the fact that the fossil fuel industries continue to funnel hundreds of billions of dollars into the production and expansion of coal, oil, and gas (2021).[144]

But it doesn't have to be this way. The many different parts of the energy grid have always been evolving. So, we can swap any part of the electricity grid like pieces of a large puzzle. What this means is that the main thing standing in the way of producing lower-emissions energy is the choice to make the change right now. We can decide collectively, and the marketplace will respond. When we switch our energy use to reduce emissions, industry and the economy will follow.

High-income countries can also choose to assist developing nations to expand access to electricity using renewables. This supports the Sustainable Development Goals because a well-established energy system supports other sectors, from business, to agriculture, transport, healthcare, and education.[145]

We can no longer accept "business as usual" because this will lead to increasing temperatures and worsening impacts across the globe. To solve the climate crisis, we must change the way the energy industry does business to make it sustainable and to achieve zero emissions. In this section we will look at how we can reduce our energy-related emissions, share our successes, and spread the word to encourage others. We will look at a range of unconventional ways to influence and compel energy providers and government to phase out fossil fuels now and replace them with renewables.

Objectives

These support the main objective to reduce emissions, as well as additional UN Sustainable Development Goals related to energy:

- Encourage and support the energy industry to reduce emissions by 45% by 2030, then to net-zero as quickly as possible.[146] This includes all energy retailers and technology and infrastructure providers.
- Ensure universal access to energy. Increase substantially the share of renewable energy in the global energy mix and double global energy efficiency (SDG 7.1, 7.2, 7.3).
- Enhance international cooperation and access to clean energy research and technology. Expand infrastructure and upgrade technology for supplying sustainable energy in developing countries (SDG 7a, 7b).
- Rationalize inefficient fossil-fuel subsidies that encourage wasteful consumption by removing market distortions (SDG 12c).
- Indigenous lands should be protected from harm by oil, gas, and mining companies in accordance with the United Nations Declaration on the Rights of Indigenous Peoples.[147]

These are some of the interconnected solutions we need to create a better future. You can find out more about the United Nations Sustainable Development Goals at **https://sdgs.un.org/**

How to achieve the objectives

In 2017, an international team of researchers confirmed that national electricity grids based on 100% renewable energy "are not only feasible, but already economically viable and decreasing in cost every year."[148] Many examples also exist around the world where large grid-scale battery installations are helping to avert blackouts and lower costs, saving consumers money.[149] Many countries already produce more than 80% of their electricity from renewable energy, as you can see from the table below.

Electricity Output: Percentage Renewable

100%	Albania, Lesotho, Nepal, and Paraguay
90% to 99%	Bhutan, Central African Republic, Costa Rica, Ethiopia, Dem. Rep. Congo, Iceland, Liechtenstein, Malawi, Namibia, Norway, Tajikistan, Uganda, and Zambia
80% to 89%	Afghanistan, Andorra, Burundi, Greenland, Kenya, Kyrgyz Republic, Lao PDR, Mozambique, New Zealand, and Uruguay

Source: International Energy Agency (IEA) and OECD via the World Bank[150]

If small developing countries can produce more than 80% of their energy from renewable sources, then developed and high-income countries should also be able to significantly improve their energy mix. Some countries have also set clear objectives to increase their share of renewable energy by more than 50% before 2030.[151]

One thing that many of these successful countries have in common is that their leaders set ambitious goals for renewable energy generation and supported them with effective policies and incentives. We should recognize that there is no one-size-fits-all solution. Recognizing this, researchers from Stanford and Berkeley universities plotted out roadmaps for 143 nations to transition to 100% renewables by 2050.[152]

They projected that in the United States, this would create 2.8 million new jobs, as seen on the diagram below.

100% UNITED STATES

Transition to 100% wind, water, and solar
(electricity, transportation, heating/cooling, industry)

Residential rooftop solar
8%

Solar plant
25%

Concentrated solar plant
7.3%

Onshore wind
30.9%

Offshore wind
17.5%

2050
PROJECTED ENERGY MIX

Commercial / govt rooftop solar
7.4%

Wave energy
0.4%

Geothermal energy
0.5%

Hydroelectric
3%

Tidal turbine
0%

40-Year Jobs Created
Number of jobs where a person is employed for 40 consecutive years

Operational jobs created: **2,815,850**
Construction jobs created: **2,285,816**

Using renewable electricity for everything, instead of burning fuel, and improving energy efficiency means you need much less energy

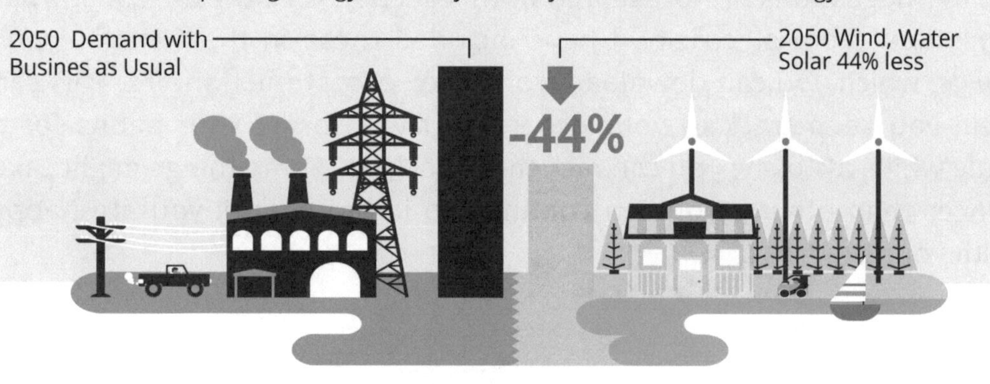

Source: The Solutions Project, Joule, and Stanford University[153]

Personal Actions

We have many options for reducing energy use, so I've organized them into several areas. If your country has a 100% renewable energy grid, then you don't have to reduce your energy use. However, for people living in countries that are transitioning, we should try to find opportunities to make a difference every time we use energy.

Getting the right numbers

If you are going to compare the cost of switching to a renewable energy provider or get a quote to install solar panels, it will help to know how much energy you are using. One way to do this for electricity and gas is to check your bills. Electricity is measured in kilowatt hours (kWh), while natural gas (methane) is often measured in British thermal units (BTU), in megajoules (MJ), or in units of 100 cubic feet (Ccf) in the United States.[154] You should be able to get the average daily usage for each month. Some providers have an online or mobile app you can use to get information, or you can also use an energy monitor.

Here to help

Many suggestions are presented in this section for how to reduce your emissions from electricity. I have included these in the *Climate Action Guide*, which you can download from **www.climate-action.org**. This can help you keep track of your progress. If you forget your habits for a little while, it's okay; you can pick them back up. Some things might take longer than others. You can continue to improve until you are happy with your progress.

Review energy use, options, and plans

Several ways are available to reduce or eliminate your energy emissions. If you are renting, you can skip to the next section, "Switching to Renewable Energy." If you own your home, then installing solar is one option you may consider.

Installing solar

You might have the opportunity to install solar panels, which will cover your energy during the day. You can also add a storage battery, which will supply energy at night. Another option is to buy renewable energy for the times you are not generating power, which is covered in the next section.

Government subsidies or other incentives might be offered in your area. Check to see if the incentives apply to the purchase or finance of equipment. This can be done online by searching for "home solar incentives in [your location]" or by asking a local installer. The following is a process that you could use for reviewing and selecting a solar installer:

1. Look at independent reviews of solar installers and pick the top three.
2. Look at their company website and see if they seem professional. Make sure they operate in your area, check that they have ways for you to contact them, and check their warranty, servicing, and purchase or lease options.
3. Get quotes from two good operators who are licensed and consider that the cheapest option is not always the best.
4. They should also provide you with an estimated amount of savings each year. If this is divided by the total installation amount, it should be possible to find out how many years it will take to pay off the system.

To get specific information on what the options are available in your area, you can search online for "How does home solar work in [your location]?" The US Energy Department has a good explanation on their website: **www.energy.gov/eere/solar/homeowners-guide-going-solar**. They also have a detailed planning guide for implementing solar power systems.

Residential Solar in Hawaii

Source: Jeremy Bezanger on Unsplash

Switching to renewable

Before I started buying renewable energy I completed the energy-saving tips in the next few sections. After saving some money on my monthly bills, I then switched to renewable energy. This is the process I followed for reviewing renewable energy providers:

1. Look at independent reviews of the best renewable energy providers and pick the top three. You can search online for "reviews of renewable energy providers in [your location]."
2. Look at their company website and see if they look professional. Make sure they operate in your area, check that they have ways for you to contact them, and look at their renewable energy options.
3. Get quotes from the best two.

If switching to renewable energy is currently too difficult financially, then see how the energy saving tips go first. We will also look at options for offsetting emissions in the review section.

Natural gas (methane)

Some people are trying to promote natural gas as a cleaner alternative to coal. However, natural gas is made up of 70%–90% methane, and when burned it emits carbon dioxide (CO2).[155] Natural gas (methane) absorbs 30 times more energy from the sun than carbon dioxide, which makes it a significant part of the problem.[156] This is an issue because there is methane leakage at nearly every point in the extraction, processing, and transportation process.[157]

If you use gas for cooking, water, or space heating, consider switching to electric if you can. Electricity is also safer, because gas heaters can leak carbon monoxide, which kills more than 430 people and hospitalizes more than 50,000 people in the United States every year.[158]

Cooking with gas also makes nitrogen dioxide (NO2), which the EPA has found is linked to an increased risk of asthma in children, diabetes, poorer birth outcomes, cancer, and premature death.[159] If natural gas is mostly methane, would it be more honest to call it methane gas instead? New homes and businesses should not be connected to methane gas. Methane should be phased out from industry, business, and homes as soon as possible.

Gas—Not the Best Alternative

Source: Kwon Junho on Unsplash

PERSONAL ACTION REVIEW ENERGY USE, OPTIONS, AND PLAN

Review your renewable energy options and select the one right for you.

 STEP 1: Switch away from natural gas for heating and cooking

STEP 2: Choose renewable option

Option A	Option B	Option C	Option D
Install solar panels and battery	Install solar and use renewable energy	Use renewable energy	Offset emissions

+

 STEP 3: Write out the steps to make this happen

 STEP 4: If the plan is complicated, keep track of your progress

What I did

My father and I looked at the alternatives for solar and battery systems. We reviewed different installers and saw a sizeable difference in price. I wasn't sure how important quality was, so I rang a friend who runs a solar installation company that specializes in large systems for business. He told me that about 10% of his business was removing cheap solar installations that had failed.

He suggested using a company that had been around at least a few years, so that if anything went wrong in the future, they would still be around to fix it. We looked at the reviews of several solar panel installation companies, and then requested two quotes. The overall cost was reduced by government incentives which made converting to solar much more affordable. Both providers had payment plans which also helped. Unfortunately, we have had a few setbacks. We were told by the installers that we would have to make some repairs to the roof and trim a few trees. Hopefully, we will have the repairs completed and the solar panels installed by summer.

Consider Residential Solar

Source: Moisseyev on iStock

Reduce emissions in the home

Heating and cooling are estimated to make up 44% of energy use in the home in the United States (2020).[160] Reducing energy use will not only reduce emissions, but in many cases will save money as well.

PERSONAL ACTION REDUCE EMISSIONS FROM HEATING AND COOLING

Smaller one-off tasks

- Try turning the water heater down a quarter. If you don't notice the difference, try turning it down again until you find the right level.
- In the bathroom, install a water-efficient showerhead.
- Reduce gaps under doors and around windows and skylights. Use block-out blinds, curtains, or window insulation film.
- Consider a smart thermostat, which can help fine-tune your usage.

Healthy planet habits

- Try turning down the heating a little in winter (20°C / 68°F). In summer, try to use fans before the air conditioner, and adjust it slightly higher in summer (25°C / 75°F); this can save around 10%.[161]
- Only heat or cool rooms being used and close doors and central heating ducts to rooms not being used. Use windows when the temperature changes instead of relying on heating and cooling.
- If you are going away, think about turning your water heater down.

Bigger one-off tasks

- Insulation can make a big difference in saving money on heating or cooling your home and reducing your emissions. Prioritize the roof.
- When purchasing a water or space heater, consider buying electric. These are also safer than kerosene and gas heaters.[162]

PERSONAL ACTION REDUCE ENERGY USE IN THE LAUNDRY AND KITCHEN

Smaller one-off tasks

- Check to see if your fridge has 1–2 inches space at the back to allow for good air circulation.

Healthy planet habits

- Try to use your dishwasher when there is a full load and, where possible, use a short or economy cycle.
- For clothes washing, wait for a full load. A lot of energy for washing clothes comes from heating the water, so consider trying cold water washing for some items. Many items can be washed effectively with detergents specially formulated for cold water. If you prefer to use warm water, try setting it to (32°C / 90°F).
- Where possible, hang clothes out to dry on an outside line, or an indoor drying rack.
- Ensure fridge door seals are tight with no gaps, so cold air doesn't escape. If you have a second fridge, only turn it on when you need it.
- Clean air filters for the clothes dryer, heaters, and air-conditioners.

Bigger one-off tasks

- When purchasing appliances, look for the energy rating and try to buy items with a better rating. See the energy rating label examples on the next page. Water efficient appliances use less hot water, which means less energy for heating. Consider replacing a gas stove with an electric induction stove.

Examples of Energy Rating Labels from the United States and Europe

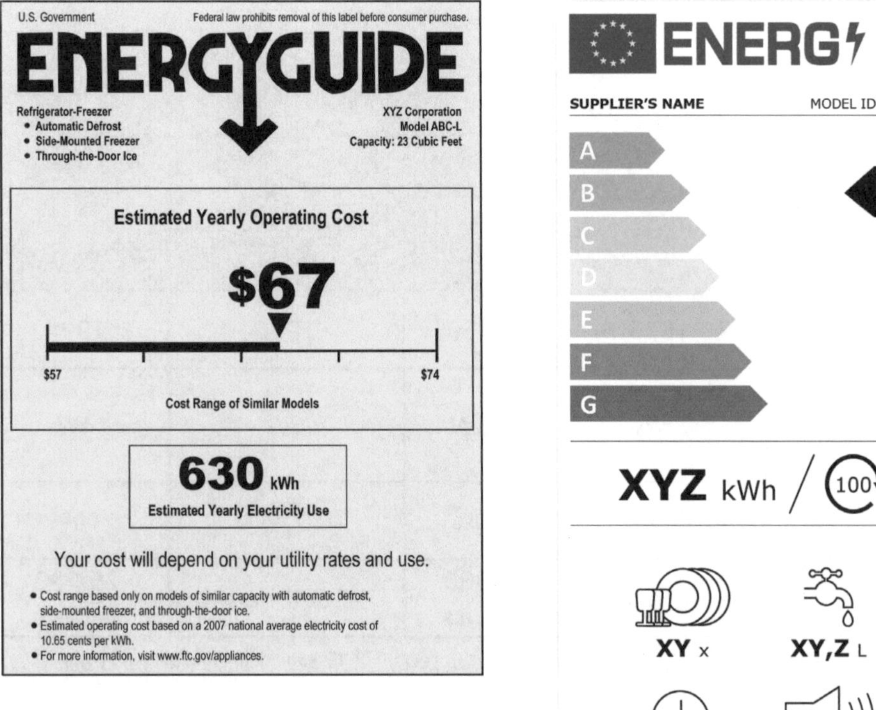

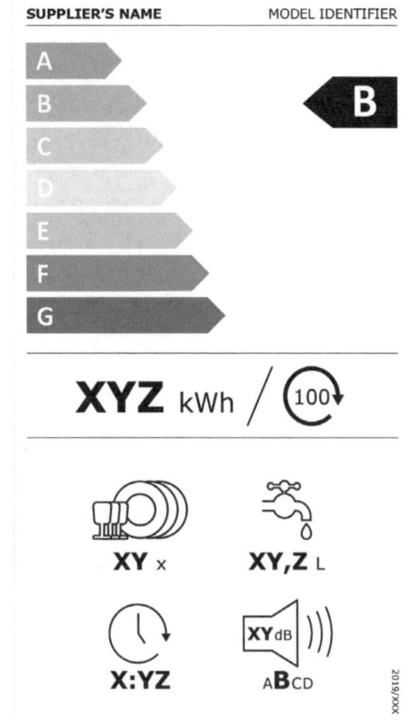

Source: US Federal Trade Commission and the European Commission

Light Bulb Comparisons

EVOLUTION OF THE LIGHT BULB	 INCANDESCENT	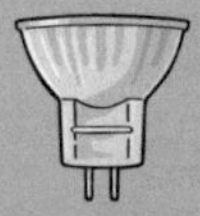 HALOGEN	 COMPACT FLUORESCENT (CFL)	 LIGHT EMITTING DIODE (LED)
Year Developed	1879	1959	1976	1994
Energy Needed for 250 Lumens	25W	18W	6W	4W
Energy Lost to Heat	90%	80%	50%	10%
Average Life Span	1,000 hours	2,000 hours	10,000 hours	25,000 hours

Source: United Nations Food and Agriculture Organization (FAO);[163] *Images: Den Potisev on iStock*

Some appliances use a lot of energy when in standby mode; this can account for up to 10% of residential energy use.[164]

PERSONAL ACTION REDUCE ENERGY USE IN OTHER AREAS

Smaller one-off tasks

- Install energy-efficient light bulbs, such as LEDs. You can see the difference your choice of light bulb makes from the table on the previous page.

Healthy planet habits

- If you're not using an appliance, turn it off. The worst offenders are TV, recorder/players, game consoles, cable boxes, stereos, computers, screens, laptops, and printers. As often as you can, turn them off at the power socket or plug them into a power strip and turn the power strip off when not in use. Another alternative is a smart plug, which you can program to turn on/off at specific times.
- Unplug rechargers for phones, laptops, tablets, or turn off at the switch when not in use.
- Turn off lights in rooms you are not using.
- Don't leave a laptop or computer running when not in use. Some computer parts will last longer if they are used less.

Bigger one-off tasks

- If you have a pool, look at using a cover to keep the heat in and install an efficient filter pump; also consider solar for heating.

Communicate

COMMUNICATE SHARE INFORMATION ABOUT YOUR EXPERIENCES

Start with someone you know

Let people know how much you have reduced your emissions and if you are saving money. You could tell a story about how easy it was to reduce your energy usage; or if it took a little while to get into new habits, you could say you got there in the end. Don't forget to share your successes, tips, and tricks with friends and family on social media.

Creating Momentum

If you feel that this book is useful in reducing emissions and creating positive change, then you can help create more momentum by letting other people know about it. You could show people a copy of the book or mention it on social media. There are pictures of the book you can download from **www.climate-action.org** and easily share.

Connect

CONNECT DISCUSS REDUCING ENERGY USE

Face-to-face

Start your own group of your family and friends to go through the steps for energy saving. You could also see if your workplace is reducing their energy use or switching to renewable energy. This can improve efficiency and reduce costs.

CONNECT JOIN CLIMATE ACTION GROUPS

You can join climate action groups by searching online for one in your area or from the following list:

- **350:** International movement of people working to end the age of fossil fuels and build a world of community-led renewable energy: **http://www.350.org**
- **Extinction Rebellion:** International movement using nonviolent direct action and civil disobedience: **www.rebellion.global**
- **Fridays for Future:** Youth-led movement that started after Greta Thunberg began a school strike for climate: **www.fridaysforfuture.org**
- Friends of the Earth: Grassroots environmental network uniting activist groups on every continent: **www.foei.org**
- **Greenpeace:** Global campaigning network across Europe, the Americas, Africa, Asia, and the Pacific: **www.greenpeace.org/global/**
- **Indigenous Environmental Network:** Grassroots Indigenous peoples who address environmental and economic justice issues: **www.ienearth.org**
- **Sunrise Movement:** Youth movement in America to stop climate change and create millions of jobs in the process: **www.sunrisemovement.org**
- **World Wildlife Fund:** Collaborates with people around the world to deliver solutions that protect communities and wildlife: **www.worldwildlife.org**

What I did

I volunteered with Fridays for Future on several protest marches as a marshal. These marches were featured in the international news and helped to raise awareness. I also joined Greenpeace and took part in several climate actions. Now more than ever we need to support climate action groups. We can donate money to help them run protests and actions (some have memberships); we can also join in and support them with our time. Many of these organizations have different ways you can help, from participating in protest actions to supporting social media campaigns. In many cases you can choose your level of involvement and be trained and supported. I was made to feel welcome and felt part of a positive community.

Influence

We have many opportunities to make a difference in how energy is generated and how quickly we can reduce emissions. You can work through them all or select the ones that are most relevant or interesting to you.

Encourage your energy provider

A report titled "The Growing Power of Consumers" says that "consumers have recognized the power behind collectively influencing the products or services they buy. Empowered consumers are actively sharing their views, and as a result are becoming more involved in the development of products and services."[165]

INFLUENCE — ENCOURAGE YOUR ENERGY PROVIDER TO HAVE RENEWABLE OPTIONS

Your energy provider

If your energy provider does not have an option for 100% renewable energy, then you can contact them and ask them to make one available. Let them know that you intend to switch to another provider if they don't have an option soon. You could do this by phone, email, a contact form on their website or through the energy company social media.

Source: Bombermoon on iStock

Government leadership

INFLUENCE LOBBY GOVERNMENT FOR ENERGY EMISSIONS REDUCTION PLAN

The state or national government

Find out if your country has reached 100% renewable energy. If not, then the national government should have a plan that will make this happen. You can search online: "Has [state/country name] reached 100% renewable energy?"

Lobby the government

Find out who is the state or national secretary or minister for energy; they might have different titles depending on where you are. You might meet with a state or national politician or contact them by phone, email, or social media. You could do this on your own, with friends, or by finding a local political or climate action group online.

Make it your own

The outline to lobby government for this section is in Appendix E: Framework for Energy. You can change this to make it your own and then discuss it or send the plan to your local political representative. The plan includes supporting developers of renewable energy to start projects, connecting them to the grid, incentivizing businesses, and households to install solar, promoting energy efficiency, and many other activities.

What I did

I met with a national politician, and we discussed the "Government Plan for Energy." They listened, smiled politely, and said they would take my notes back to their people. Before leaving, I promised I would only vote for someone who was serious about action on climate change. It seems that for many politicians, the most important priority for them is to get elected. If enough people tell them that action on climate change is important, then it might create enough concern for them to listen and do something meaningful. In the recent election, I voted for an independent and they won!

Compel

No new coal

A good way to start fixing a problem is to stop making the situation worse. At a climate summit in 2019, the UN Secretary-General António Guterres called for no new coal plants to be built from 2020.[166] However, there were more than 200 new coal-fired power stations under construction and more than 300 in preconstruction in more than 30 countries in 2021.[167] Many of these new coal-fired power station projects are in developing countries. These countries have the right to provide electricity to their people at the lowest cost possible. To help avoid an increase in emissions, the international community should work with developing countries and provide them with technological assistance and funding to help give them the option to deploy renewable energy instead.

Coal-Fired Power Stations – Should be a Relic of the Past

Source: Fotoforce on iStock (smokestacks on left and steam rising from cooling towers on right)

COMPEL STOP NEW COAL-FIRED POWER STATIONS

Find out

See if any new coal-fired power stations are planned to be built in your country. The Global Energy Monitor provides a tracker at **www.globalenergymonitor.org/projects/global-coal-plant-tracker/tracker/**. The map lists "coal-fired units" (note that several units are often located at each power plant). In the right corner in the box titled "Coal plants: Status," uncheck all boxes except "Pre-permit," "Announced," and "Permitted." These are the power stations that might be stopped from being built. Citizens of countries with new coal-fired projects in development can encourage their government to look at renewable options. If none are planned in your country, skip to the next section.

Find groups already in action

You can zoom in on each plant listed and find out information about the location, status, and name of the power station. Do an online search to see if construction is still going ahead and if any groups oppose it. You can join and support an existing group. If not, see if any climate action or environmental groups are nearby and if they want to act.

Identify and pressure local contractors, state, and national government

You could put pressure on local and state governments or find out which organizations are involved in building the power station. These can be engineering and construction firms, insurance, and transportation, as well as banks providing finance. One action could be to publicly name and shame them.

Close the dirtiest power stations

Coal has the biggest share of global electricity generation, but not all coal-fired power stations are the same.[168] Researchers have found that the worst polluting power stations are responsible for most emissions.[169] This knowledge offers us a great opportunity to make a positive difference in the shortest period. If the top 5% of these hyper-polluting power stations were shut down, then up to 49% of the world's emissions from electricity generation could be eliminated.[170] This action would substantially help achieve the goal of reducing emissions by 45% by 2030. A staged shutdown of the worst coal-fired power stations first means there will still be enough energy to provide grid stability, as the share of renewable energy increases.

COMPEL CLOSE HYPER-POLLUTING POWER STATIONS

Just start with one!

I believe that closing the worst hyper-polluting coal-fired power stations is one of the best opportunities to turn the situation around and finally reduce emissions.

All developed and high-income countries that have coal-fired power stations should shut down one of the worst hyperpolluting plants in their country and replace it with renewable energy.

On the next page is a process called "5 Steps: Switching to Renewable Energy". You could discuss this with your national politicians and promote it to people you know personally and on social media. You can download the infographic from **www.climate-action.org**.

5 STEPS

SWITCHING TO RENEWABLE ENERGY

If we shut down the top 5% of the worst polluting power stations, then up to half of the world's emissions from electricity generation could be eliminated. The objective of this plan is to help reduce emissions by 45% by 2030.

1 Engagement

Coal-fired power station operators volunteer to join a transition program led by the government. Each operator nominates the highest polluting power station to be closed and commits to building a replacement using renewable energy. The local community is consulted, with government providing funds for training, supporting new industries and job creation.

2 Funding

Many low-cost financial options, such as green bonds, could be used to fund the cost of new renewable energy development. These could be backed by the government.

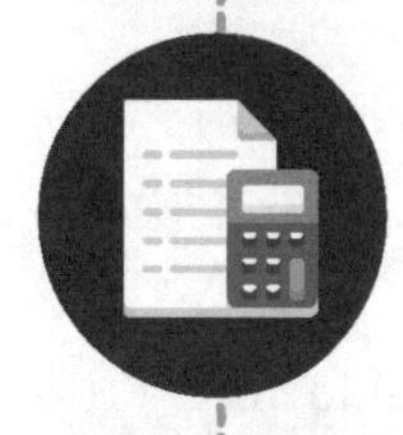

3 Profitability

The government provides subsidies and tax incentives to participating energy operators for an agreed-upon term. This will also promote price stability for the community. Subsidies for providers refusing to participate can be phased out.

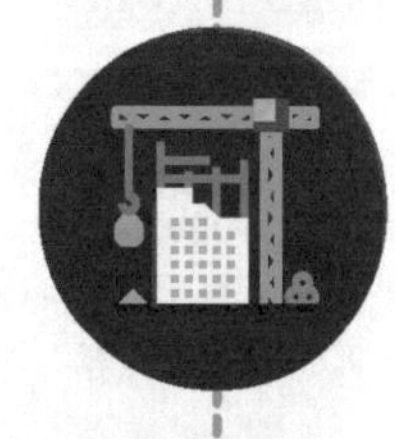

4 Building

Renewable power with a similar energy capacity is built through consultation with the local community. The best options and locations for renewable energy should be selected, along with connectivity to the network.

5 Transition

After the new renewable power comes online, the existing coal-fired power station is decommissioned. With this process proven to work, more coal-fired power stations can be transitioned to renewable energy.

Energy Conclusion

One thing is clear: continuing the way things are now will lead to higher temperatures across the entire globe, affecting all people. Therefore, business as usual should be considered the enemy of all life on our planet. We cannot wait for politicians who take money from fossil fuel corporations or for industries that don't want to change. We cannot wait another 100 years for renewable energy to replace fossil fuels. To solve the climate crisis, we must change the way the energy industry does business today.

Many small and developing countries have already demonstrated that switching to renewable energy is possible. We need to demand that our leaders set ambitious goals for renewable energy. Many solutions are available, and we can start changing and making a much better future for ourselves and generations that follow, but we must start now. As consumers, we can change the way we use electricity, and this can influence the direction of the energy industry. We can communicate and connect with other people to share our successes in reducing emissions and our positive vision for the future. We can influence electricity suppliers and government, especially when we work together.

Review your progress and see what emissions reductions you have made and recognize your success. You can take time out and celebrate with people you have been reducing emissions with, or with your accountability buddy, friends, or family.

Transportation

Source: Bithin Raj on Unsplash

ACTIONS – TRANSPORTATION

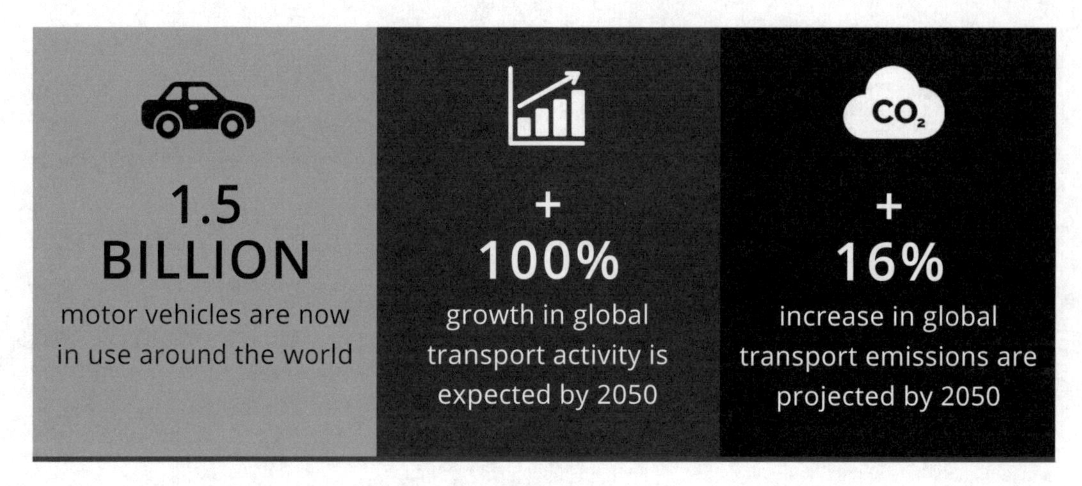

Sources: International Transport Forum (ITF) and Global Fuel Economy Initiative (GFEI)[171]

A great migration of people has meant that half of the world's population now live in cities.[172] This trend is expected to continue so that 70% of people in the world will live in urban areas by 2050.[173] This rapid change is increasing transportation activity and emissions. But it doesn't have to be this way. With people coming to live in the same area there are big opportunities for positive change.

Situation

As more and more people move around cities, interstate, and overseas, our relationship to transportation becomes an increasingly significant part of the emissions we produce. On current pathways, passenger and freight transportation are both expected to increase by 100% by 2050.[174] At an international summit of transportation ministers in 2021, it was reported that even if the current limited commitments to decarbonize transportation were fully implemented, emissions from the sector are expected to rise by 16%.[175]

Some people say that technology will save us from the climate crisis, but I'm not so sure that we can sit back and wait for that to happen. The first electric cars were being sold commercially in the 1890s, and by 1901, more than a third of cars in the United States were electric.[176] This technology failed to progress, and even after the international community agreed on climate action in the 1990s, very little was done to transition to clean transportation. The slow development of electric vehicles meant that in 2023, they made up only about 3% of the 1.5 billion vehicles on the road.[177] Mainly due to new entrants into the market, the industry has finally begun mass-producing electric and other zero-emissions vehicles. However, many other sectors of transportation, such as air and sea, have been slow to develop solutions.[178] We cannot wait another 100 years for technology to slowly replace old combustion engines.

Carbon Footprint Per Passenger in Kilometers

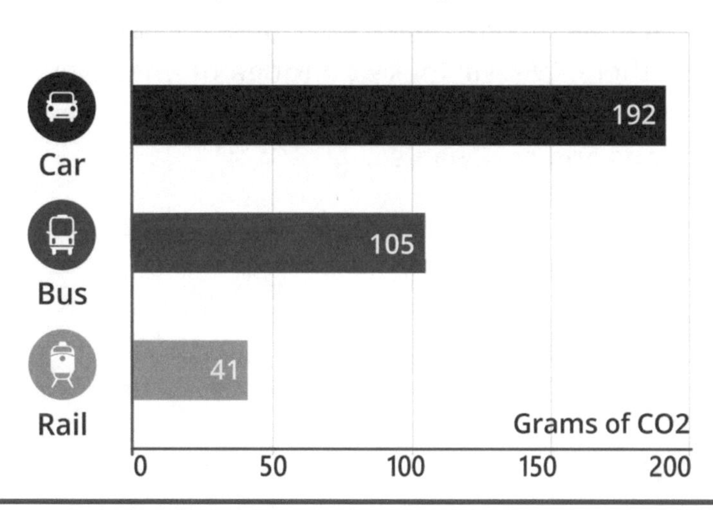

Source: UK government via Our World in Data

Another issue is the inefficient use of resources required to move only one person from one location to another. One person traveling in a car causes almost twice the emissions of a passenger on a bus, and four times that of rail over the same distance.[179]

So many opportunities are available. National governments can encourage zero-emissions passenger and freight options. With their high density of people, cities are also in an ideal position to reduce emissions from urban transportation by 80%, according to the International Transport Forum.[180] They can promote and expand public transport, improve bike paths, and provide other mobility options.

Some people can afford to purchase zero-emissions vehicles, while others may have options to share a ride to work, take public transport, or use bicycle paths. When we share transportation solutions, we consolidate the act of moving people from one place to another. Consolidating our transportation choices will lead to low emissions.

To solve the climate crisis, we must change the way the transportation industry operates in order to make it sustainable and to achieve zero emissions. In this section, we will see how our choices can reduce our emissions, as well as how to share our successes and spread the word to encourage others. We will look at a range of unconventional ways to influence transportation providers and government to implement zero-emissions options and renewables.

Objectives

These objectives support reducing emissions, as well as the UN Sustainable Development Goals that relate to transportation:

- Support and encourage the transportation industry to reduce their emissions by 45% by 2030, then to net-zero as quickly as possible.
- By 2030, provide access to safe, affordable, accessible, and sustainable transportation systems for all, improving road safety, notably by expanding public transport, with special attention to the needs of those in vulnerable situations (SDG 11.2).
- Integrate climate change education and measures into national policies, and planning (SDG 13.2, 13.3).

These are some solutions we need to create a better future. You can find out more about the Sustainable Development Goals at **https://sdgs.un.org/**

How to achieve the objectives

Several connected solutions are required to improve transportation systems:

1. Consumers change to a lower-emissions transportation lifestyle.
2. Transportation manufacturing industries create affordable options for consumers, mass transportation, and freight.
3. Government creates policies to support consumers and industry to use zero-emissions options, as well as to expand public transportation infrastructure.
4. Industry and government invest more in research to solve technically difficult solutions to lower transportation emissions.

We will look at a range of options for how to reduce our transportation emissions, then communicate our progress to help and inspire others. There will also be a range of ways to influence the government to create the right conditions for rapid implementation of renewable transportation solutions.

The good news is that many of the major car manufacturers now have zero-emissions models for sale. Several of the biggest names have even set an end date for phasing out old technology internal combustion engines.[181] The inevitable movement toward zero emissions is on the way; we just need to create greater acceleration.

Personal Actions

Vehicle usage (motorbike, car, SUV, or van)

I would like to buy an electric vehicle, but I can't afford it now. However, I plan on making my next car purchase an electric one. I found I could make many changes to the way I use transportation to reduce emissions.

An Electric Vehicle in a Designated Charging Bay

Source: Andrew Roberts on Unsplash

The following tips are mainly for old technology internal combustion engines, but many can also help for zero- or low-emissions vehicles to reduce energy use and increase range.

PERSONAL ACTION LOWER EMISSIONS FROM VEHICLE USAGE

Smaller one-off tasks

- Reduce weight by removing unnecessary items from your car.

Happy planet habits

- Remove accessories that increase wind resistance, like roof racks, when not in use, as they can decrease miles per gallon (about 17% for fuel cars).[182]
- Regularly check tire pressure. Properly inflated tires last longer and increase miles per gallon.
- Keep the temperature setting at the lowest for heating and highest for cooling that will make you comfortable. Parking in the shade on a hot day can help to reduce the amount of cooling you need.
- Aggressive driving—speeding, rapid acceleration, and braking—can lower your mileage (for fuel cars by about 10% to 40% in traffic).[183]
- Mileage improves by observing the speed limit, so try cruise control on freeways if you have it to save about 7% to 14% for fuel cars.[184]
- For fuel cars, use the recommended oil for your vehicle and keep it serviced and tuned regularly.

Bigger one-off tasks

- Next time you buy a vehicle, consider an electric or hybrid. If you can't afford one, then think about a car with a more efficient engine. Information on fuel efficiency is available at www.fueleconomy.gov.

Reducing emissions promotes health

Currently, 90% of the global population breathes unsafe air, exceeding recommended limits for outdoor air pollution, according to the World Health Organization.[185] This is a problem because air pollution increases the risk of cardiovascular disease, as shown in the table below.[186]

Air Pollution as a Cause of Cardiovascular Disease

24% of deaths from **STROKE**	25% of deaths from **HEART DISEASE**	43% of deaths from **LUNG DISEASE**

Source: World Heart Federation

As we move toward lower emissions, pollution from motorcycles, cars, trucks, trains, and airports will decrease. Our cities will have cleaner air, which means reduced impact from air pollution.

Imagined Cityscape Moving from Polluted to Cleaner Cities

Source: Alxey Pnferov on iStock

PERSONAL ACTION TRAVEL LESS OR CHANGE THE WAY TO TRAVEL

Less travel

Try having fewer trips through planning. One way could be to plan your shopping more efficiently and go less often or to combine shopping for several things in one trip.

People power

If you are not going long distances, look at walking, using a scooter, or riding a bike. The American Heart Association advises that being more active can lower the risk of heart disease, stroke, Type 2 diabetes, high blood pressure, dementia, Alzheimer's, and several types of cancer.[187]

Holidays

If you plan a holiday, consider traveling inside your own country or your local region and look at lower-emissions options.

Commuting to work

While many people don't have the option to work from home, it might be possible for some people. For those whose job does not require them to be at the workplace, this possibility could be discussed with your supervisor. It might not be possible to work all days from home, but maybe one or two. If you drive to work, to avoid the peak rush times, you might start early and finish early (or start late and finish late). This might reduce the time you spend in traffic, reduce your emissions, and save money.

Public transport

See if you can use public transportation more often. If you don't have enough public transportation options in your area, then consider adding this to the list of things to ask your local or state politicians.

Communicate

The International Energy Agency (IEA) developed a plan titled "Net Zero by 2050."[188] It calls for "a rapid shift away from fossil fuels" and sets out 400 milestones to avert the worst effects of climate change. One milestone in the plan sets a date for the end of sales of old-technology internal combustion engine passenger cars by 2035.[189]

COMMUNICATE COMMUNICATE WITH MANUFACTURERS

Communicate to car and motorbike makers

Do the car or motorbike manufacturers in your country have several zero-emissions vehicles for sale? Have they set an end date for making the old internal combustion engines by 2035? If not, then you can send them a message:

- An email through the manufacturer's website
- A post, comment, or question on their social media sites
- A complaint through their customer service
- A product review that asks for climate action now
- An email or comment on social media to car dealerships
- Posts on your own social media
- Letter to the editor of the local newspaper

The age of the old internal combustion engine is over.

Sharing your experience and success

Tell people how you are communicating with manufacturers and how you are reducing your transportation emissions.

Connect

In addition to individual communication, you can establish broader connections.

 CONNECT — CONNECT WITH PEOPLE TO REDUCE TRANSPORTATION EMISSIONS

Sharing and consolidation

You can reduce your emissions by sharing transportation with others, it also saves maintenance costs as well as wear on tires:

- Carpool or use ride-share programs with other people at your work or people from your area who are going in the same direction.
- Join neighbors or friends for shopping or to pick up children from school.

Cycling together

If you like riding a bicycle, then think about how you can incorporate more cycling into your lifestyle. Find out if riding to work is practical and see if anyone will ride with you. You can also see if there is a local cycling club. They might lobby local or state government for better bicycle tracks and lanes to improve rider safety and travel time.

Influence

Two areas in the transportation industry need immediate action. The first is the rapid deployment of technologies that are already available, such as electric and hydrogen options. The second is the quick development of the technology for other modes of transportation that require solutions, such as shipping and air travel.

Promote zero-emissions technology available today

The International Energy Agency (IEA) plan "Net Zero by 2050" asserts that a substantial deployment of clean energy solutions must happen between now and 2030.[190] This means that the sales of hybrid, electric, and hydrogen vehicles must increase quickly, as outlined in the table below.[191]

Sales of Low or Zero-Emissions Vehicles Required by the IEA Plan

TYPE OF VEHICLE	PERCENTAGE OF GLOBAL SALES BY 2030
2 and 3 wheelers	85%
Cars	64%
Vans	72%
Buses	60%
Heavy trucks	30%

Source: International Energy Agency (IEA) plan Net Zero by 2050.

Develop low- or zero-emissions solutions quickly

The "Net Zero by 2050" plan makes it clear that we need to achieve major leaps in clean energy innovation this decade to bring these new technologies to market in time.[192] All areas of the transportation industry which do not have zero emissions solutions ready to go now should make significant efforts to develop and deploy new technology as soon as possible. To make this happen, consumers, industry, and governments at all levels will need to take significant action from this point onward.

Government leadership

INFLUENCE — **LOBBY GOVERNMENT FOR A TRANSPORT EMISSIONS REDUCTION PLAN**

City, state, and national

Different parts of transportation policy may be organized by different levels of government in each country or region. Your city government could be responsible for bicycle paths, for example. The state or national government could be responsible for public transportation. Identify which part of government you need to contact.

Single issue

You could take a single issue from the framework to lobby government, which is located in Appendix F: Framework for Transportation. This approach could work well to campaign for more bicycle tracks or getting local or state government to switch their car fleets to zero-emissions vehicles.

Make it your own

You can edit the plan and make it your own. Then engage with state or national governments and start a conversation with them.

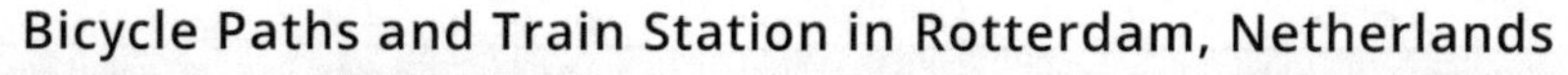

Bicycle Paths and Train Station in Rotterdam, Netherlands

Source: Jurriaan on Unsplash

Compel

Transport-related emissions from tourism represented 8% of all human made emissions in 2024.[193]

Inspiring transportation industries that are slow to change

The World Tourism Organization reported that most experts see international tourism returning to normal levels in 2024.[194] This means more than 1.4 billion international trips for tourism per year.[195] These trips for holidays are the choice of the consumer, and consumer power can be wielded to create positive change in the world. Why specifically tourism? Tourism includes travel by land, sea, and air. A container ship operator might not care what the average person in the street thinks, but a passenger ship whose entire income depends on the decisions of consumers will. New zero-emissions technology developed for passenger trains, ships, and planes can be applied to freight as well.

The industries listed in this section have already had 30 years to invest and develop solutions. When these industries commit proper resources and when they come up with the solutions will largely depend on pressure from consumers, as well as government incentives and regulation. If there is too little pressure or support, it might be decades before zero-emissions solutions are developed. For example, some major commercial aircraft manufacturers have committed to producing zero-emissions aircraft by 2035, while others see this as a "longer-term solution."[196] Instead, they are relying on carbon-neutral options, such as sustainable aviation fuels (SAF).[197] SAF is made by using biofuels or taking carbon dioxide from the atmosphere or another source and processing it into aviation fuel, which can be used by existing aircraft.[198] When sustainable aviation fuel burns, carbon dioxide is made; therefore, even though SAF is carbon neutral, it should be considered as a temporary measure and not a solution. The alternative is to make zero-emissions airplanes for passenger and freight, which has already started happening, but they have not been deployed widely.[199] More pressure from consumers can speed this up.

COMPEL **INSPIRE TRANSPORTATION INDUSTRIES THAT ARE SLOW TO CHANGE**

If land, sea, or air travel operators won't change now, then we as individuals can choose not to use their services. We can use a range of different ways to let transportation operators know, including:

- When booking in person or by phone with a travel booking center, ask for zero-emissions options only.
- Contact the air, sea, or train operator by email or posting a comment on their social media.
- Make a complaint through their customer service.
- Leave a negative product review that asks for zero-emissions travel now.

We can let the transportation industry know that we will choose a better future with the purchasing decisions we make.

Land

Interstate rail or bus operators should make progress toward zero emissions, and we can tell them they need to make change now

Sea

Let cruise ship lines know you won't book with them until they provide zero-emissions travel.

Air

You could let your national airlines know you will fly less or not at all until they can provide zero-emissions travel.

Destinations

It's not only the modes of transportation that need to reduce their emissions. You can encourage hotels, resorts, theme parks, zoos, museums, art galleries, and other tourist destinations as well, whether they be local or international.

Transportation Conclusion

While some parts of the transportation industry have made good progress, such as electric vehicles, many others have achieved little in the last 30 years. We cannot continue this way any longer. Emissions must decrease quickly before 2030.

Transportation companies will probably achieve results sooner with additional incentives from governments or if inspired by regulation or carbon taxes for failing to create positive change. We need to insist that the government set ambitious goals for the transportation industry. Governments can also provide leadership and increase accessibility to walking and bicycle paths, zero-emissions public transportation, recharge and other necessary facilities. Governments and industry need to support and invest in innovation with the aim of rapid solution deployment.

One thing is for certain: transportation industries will be influenced by strong demands from consumers. If we as consumers stop paying for their services and their profits fall, then they will pay attention. To solve the climate crisis, we must change the way the transportation industry does business.

Support for Developing Countries

Source: Dmitry Malov on iStock

SUPPORT FOR DEVELOPING COUNTRIES

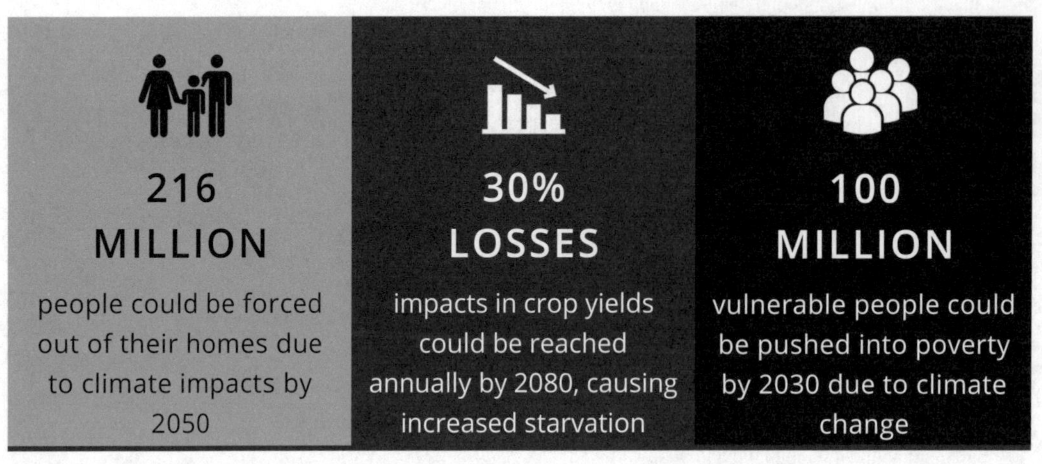

Sources: The World Bank[200]

The United Nations Framework Convention on Climate Change recognizes that the largest share of historical emissions originated in developed countries.[201] The United States and Europe account for 47% of all emissions, as seen in the following table.[202] Historical emissions are important because carbon dioxide stays in the atmosphere for between 300 and 1,000 years.[203] This means that the emissions made since the Industrial Revolution are still in the atmosphere.

Historical Share of CO_2 Emissions Between 1751 and 2017

RANK	COUNTRY	PERCENTAGE OF ALL EMISSIONS
1	United States	25%
2	European Union (28 with United Kingdom)	22%
3	China	12%
4	Russian Federation	6%
5	Japan	4%
6	Rest of the world (190 countries)	31%

Source: Our World in Data based on data from the Global Carbon Project and CDIAC.

This also applies to more recent emissions. Research by the Stockholm Environment Institute shows that half the global population earns less than $6,000 per year, and that this group is only responsible for 7% of total emissions from 1990 to 2015.[204] In comparison, the top 10% of income earners in the world are responsible for 49% of total emissions (people earning more than $38,000). The next table shows the income levels and population sizes of each of the world's income groups.[205]

Income Groups, Emissions, and Population (1990 to 2015)

INCOME	EMISSIONS	NUMBER OF PEOPLE	INCOME PER YEAR IN $US
Lowest 50%	7%	3.6 billion	Less than $6,000
Middle 40%	44%	2.9 billion	More than $6,000 & less than $38,000
Top 10%	49%	730 million	More than $38,000

Source: Stockholm Environment Institute and UN World Population Prospects 2015.

It is important to acknowledge that the lifestyles of people on low incomes have resulted in much less emissions than those of people on middle or high incomes. It should also be acknowledged that developing countries have less capacity to recover from climate disasters or to adapt to worsening impacts such as sea-level rise. In many cases, those impacts are being felt now.

Researchers estimated that 5 million people are dying every year because of climate change–caused extreme weather, with 3.8 million from Africa, South and Southeast Asia.[206] It has also long been recognized that small island nations have been responsible for a fraction of global emissions but now face the impact of climate change and sea level rise (IPCC).[207] This is an issue of climate justice, which acknowledges the fact that, in many cases, those who contributed the least carbon emissions are being most affected by climate change.

The first principle of the United Nations Framework Convention on Climate Change is that developed countries should lead the way in reducing emissions and should support developing nations based on equity.[208] Therefore, when I state in this book that we all have a role to play on climate change, please remember that the developed and high-income countries should be leading by example. This is also because developed countries have greater financial capacity to act and support developing nations.

The Green Climate Fund was established for this purpose at climate talks in 2010 (COP16). This fund is dedicated to providing financial resources to assist developing nations with reducing emissions and to adapt to climate impacts.[209] However, at the climate talks in Glasgow, it was acknowledged that developed countries had fallen short on their financial commitment."[210] After 10 years of failure, it is time to shine a light on this issue and for developed countries to fulfill their promises.

INFLUENCE LOBBY GOVERNMENT TO HELP DEVELOPING COUNTRIES

Lobby your national government

While climate agreements state it is the responsibility of "developed" countries, I would suggest that in most cases high-income countries also have the capacity to contribute, so they should be included as well. You can see the list of developed and high-income countries on the following page.

If you live in a developed or high-income country, consider lobbying your secretary of state or foreign minister to contribute to the Green Climate Fund. This money is mainly focused on assisting the least developed countries (LDCs) and small island developing states (SIDS).[211] You can find out more about the Green Climate Fund: **www.greenclimate.fund/about**.

Sustainable Development Goals

All the countries in the following table are categorized as being high-income by the United Nations Development Programme.[212] Each is ranked in order on the Human Development Index (HDI).[213] The HDI is worked out by assessing life expectancy, education levels, and gross national income per capita.[214] Take into consideration that the least developed countries (LDCs) have a gross national income per capita average of $1,273 per person.[215]

High-Income Countries with Gross National Income Per Capita

1	Norway	$66,494	21	Slovenia	$38,080
2	Ireland	$68,371	22	Republic of Korea	$43,044
3	Switzerland	$69,394	23	Luxembourg	$72,712
4	Iceland	$54,682	24	Spain	$40,975
5	Germany	$55,314	25	France	$47,173
6	Sweden	$54,508	26	Czech Republic	$38,109
7	Australia	$48,085	27	Malta	$39,555
8	Netherlands	$57,707	28	Estonia	$36,019
9	Denmark	$58,662	29	Italy	$42,776
10	Finland	$48,511	30	United Arab Emirates	$67,462
11	Singapore	$88,155	31	Greece	$30,155
12	United Kingdom	$46,071	32	Cyprus	$38,207
13	Belgium	$52,085	33	Lithuania	$35,799
14	New Zealand	$40,799	34	Poland	$31,623
15	Canada	$48,527	35	Andorra	$56,000
16	United States	$63,826	36	Latvia	$30,282
17	Austria	$56,197	37	Portugal	$33,967
18	Israel	$40,187	38	Slovakia	$32,113
19	Japan	$42,932	39	Hungary	$31,329
20	Liechtenstein	$131,032	40	Saudi Arabia	$47,495

Sources: United Nations Development Programme 2020, United Nations (2020)

PART 4

Reviewing and Updating the Plan

"Every bit of warming matters,
every year matters,
every choice matters."

—IPCC Special Report on Global Warming of 1.5°C

STEP 7. REVIEW PROGRESS

Now that we have gone through the process of reducing emissions, most people will find there will be a few areas that need more work. There are several reasons why we might have emissions left:

- In many cases, there are no zero-emissions product or service options.
- No labeling or information is given about the emissions created by products or services.
- Some changes take time, such as installing solar panels.
- It might be difficult to afford low- or zero-emissions alternatives.

Because you may have lingering emissions to work on, it will be useful to look at continuing to reduce emissions as an ongoing process.

The Process

We can approach reducing emissions as a series of steps:

1. **Plan:** Create objectives.
2. **Reduce:** Work through the actions to make personal emissions reductions.
3. **Review:** Measure the remaining emissions.
4. **Offset:** When emissions are unavoidable, offsets are activities that balance out your emissions by reducing or removing greenhouse gases from the atmosphere.[216]
5. **Net-zero:** Achieve a state of net-zero emissions when human-caused greenhouse gas emissions are balanced by removals over a specified time.[217]

We need to stop all emissions as quickly as possible and let the earth recover from the damage that has been done. However, we don't have all the solutions to stop emissions right now. If we all do the best we can, then as new solutions become mainstream, prices will continue to fall, and more people will be able to afford to reduce emissions further. Until that happens, carbon offsets are an option to achieve net-zero. Offsetting should not be used as an excuse for maintaining high carbon emissions for individuals, businesses, or governments. Some examples of offsets include supporting projects that replant forests with a biodiverse range of trees or supporting developing countries in creating renewable energy.

Carbon Footprinting

After reducing emissions, it's time to review your progress. The total amount of greenhouse gases that a country, organization, or individual is responsible for is called a carbon footprint.[218] The definition we are using for carbon footprint includes all greenhouse gases.

This means conducting a carbon footprint assessment for your household to see what emissions you have left. This gives you the opportunity to celebrate your progress, and it will also help you know the total amount of emissions to offset, if this is something you want to do.

How carbon footprint calculators work

Of the many online carbon footprint calculators, most don't cover all the areas in this action plan, such as energy, food, transport, products, and services. I found two calculators that do and that are free to use. One is The Nature Conservancy calculator, which is designed for residents in the United States. The other calculator has an international approach and is provided by an organization called Trace, which provides carbon offsets for individuals and businesses.

What I did

We worked out our carbon footprint and were surprised that it was higher than expected, because we didn't think we were overconsuming or wasteful. The good news was that we now had some useful information that would help us to reduce emissions in the future. We found that household energy and transportation were the largest contributors, so we will focus on these for the next cycle of reducing emissions.

Select the option below that suits you. Detailed explanations of how to use the online calculators are included in the *Climate Action Guide*.

PERSONAL ACTION MEASURE YOUR OWN EMISSIONS

United States - The Nature Conservatory

www.nature.org/en-us/get-involved/how-to-help/carbon-footprint-calculator/

Select a city or zip code in the United States and use the advanced options to add more information to get a more accurate estimate.

International - Trace

www.our-trace.com/tools/carbon-footprint-calculator

Select country and currency (USD = United States, GBP = Great Britain, EUR = Euro, NZD = New Zealand, AUD = Australia). If your country is not on the list, pick the closest one or select "other" and use USD. Then click "2." At the end, you are asked to enter a name and email. Regardless of what you enter, the results will still be shown on the next screen. Trace is an Australian organization, and the results page includes a native animal such as a Kangaroo or Wombat. Ignore this and record the results.

Feel free to use another carbon footprint calculator that you prefer. Other options are provided by 8 Billion Trees **www.8billiontrees.com/carbon-calculator/** and the Global Footprint Network **www.footprintcalculator.org**.

Record your emissions

Once you have calculated your carbon footprint, make a note of the total, as well as any specific information, in the *Climate Action Guide*. You can download this from **www.climate-action.org**. Also, make a note of the calculator you used. If your emissions are higher than you thought, don't be disappointed. This is just a place to start.

Offsetting and Achieving Net-Zero

The IPCC defines net-zero as a point when human-caused greenhouse gas emissions are balanced by human-caused removals over a specified time.[219] One way to balance out emissions is to financially support projects which reduce or remove greenhouse gases from the atmosphere. This can be done by buying carbon credits. One carbon credit represents the certified removal of one metric ton of carbon dioxide from the atmosphere.[220] If you have ten metric tons of remaining emissions, this could be balanced with ten credits.

Carbon Offsets: A Temporary Process

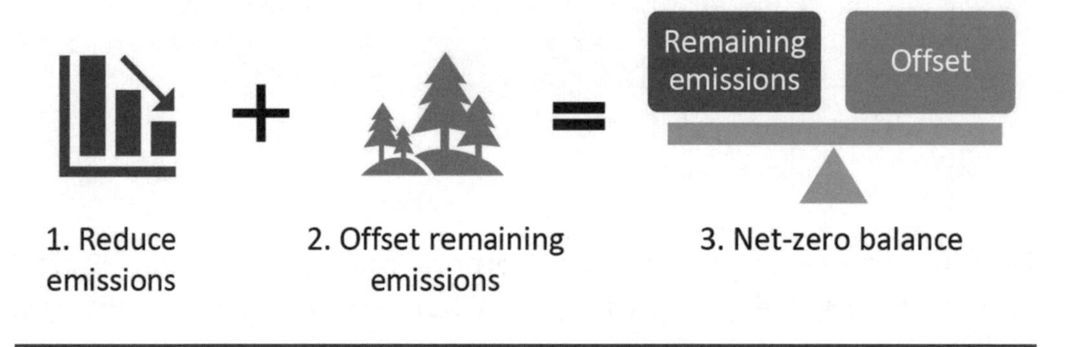

Source: British Standards Institution (BSI)[221]

Opportunities are available to support projects that reduce emissions and to help communities in developing countries on the front lines of climate change. Some examples include:

- Replanting forests with a biodiverse range of trees and native vegetation while supporting local and indigenous communities.
- Supporting developing countries to create renewable energy, which often means reducing the demand for fossil fuels or providing electricity to people for the first time.

- Many people in developing countries don't have access to energy for cooking and rely on burning wood in fireplaces. Some projects provide stoves that are 50% more efficient. This means fewer emissions, less stress on nearby forests, and less pollution in the home.

You can choose from many other types of projects.

What I did

I saved money by reducing my car usage and we have also saved on our food, energy, and consumption. We put enough of the savings to pay for carbon credits to offset our remaining emissions. I found many types of organizations offering carbon offsets online. Some projects have been criticized for not achieving what they promised, so I looked for projects that are certified by an internationally recognized organization. This certification confirms that each carbon credit represents one metric ton of carbon dioxide removal.

Two of the main certification organizations are Gold Standard and the Verified Carbon Standard (VCS) by Verra. It was also important to us that the projects aligned with our values by supporting developing countries and the United Nations Sustainable Development Goals. After reviewing several projects, we found the credits can cost from US$11 per metric ton up to US$45. Therefore, if you have ten metric tons to offset, this could cost between US$110 to US$450, depending on the projects you choose. We bought several credits from a few different projects that helped renewable energy development, reforestation, and efficient cookstoves in developing countries. These were certified by either Gold Standard or Verra.

PERSONAL ACTION — REVIEW THE OPTION FOR CARBON OFFSETTING

Identify the amount of the remaining emissions you worked out using the carbon footprint calculator. Review some offset projects, taking into consideration your financial circumstances and your values. Below are two well-known organizations who offer carbon credits which are certified.

Gold Standard

In addition to maintaining an international standard for carbon offsets, this organization has a carbon credit marketplace that has a range of carbon removal and emissions reductions projects. These also support the UN Sustainable Development Goals.

- How it works: www.goldstandard.org
- Projects to choose from: marketplace.goldstandard.org/collections/projects

South Pole

Their projects are certified by Verra or Gold Standard and have many options that are aligned to the UN Sustainable Development Goals.

- How it works: www.southpole.com/carbon-offsets-explained
- Projects to choose from: shop.southpole.com

Many carbon offset programs are in place around the world. If you are interested in using other options, search online for "certified carbon offsets in [country name]" for one that suits your region or personal objectives.

COMMUNICATE — SHARE INFORMATION ABOUT YOUR EXPERIENCES

Sharing your success

Let people know how you are doing with your emissions reductions overall and carbon offsets. This could be done face-to-face or on social media.

STEP 8. UPDATE THE PLAN

The process for addressing remaining emissions may be ongoing and be approached as a cycle, as shown below. It may take several years to achieve net-zero without offsets. Each time we go through the process, the objective is to reduce emissions so that they are smaller and smaller. You can decide how often to go through the process. One way could be to look at emissions once each year. Selecting a time of year that is normally not very busy and adding a reminder in your calendar might be helpful.

The Ongoing Process of Reducing Emissions

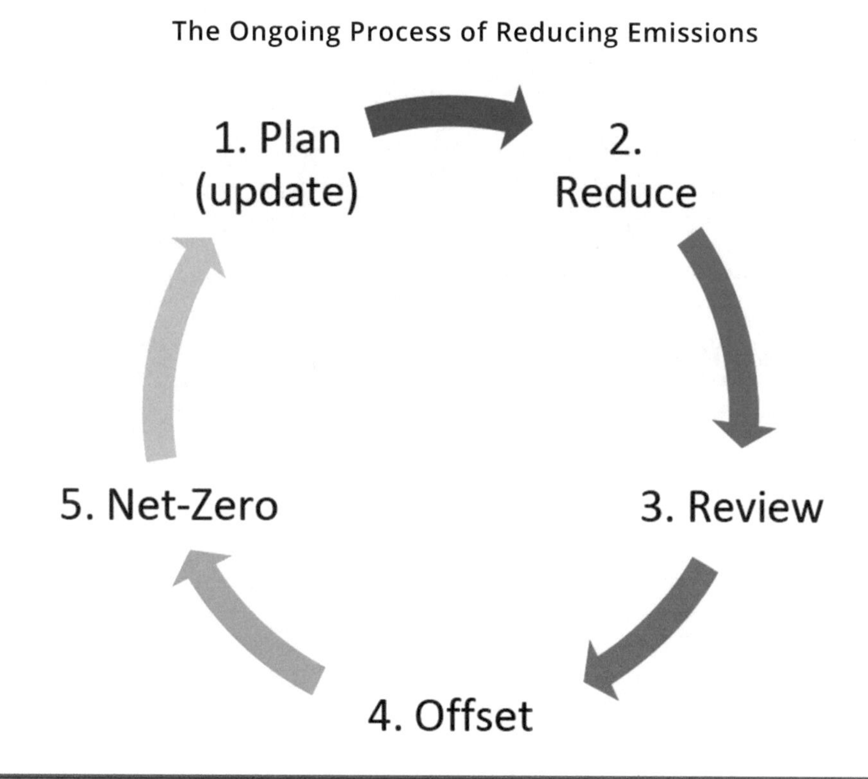

Source: Based on Climate Neutral by United Nations Framework Convention (UNFCCC)[222]

When starting the process again, think about discussing emissions reductions with your household, family, or friends. Then talk about the emissions reduction process and include them when completing your carbon footprint. The next step is to update your plan and keep the process going.

Source: lovelyday12 on iStock

Update your emissions reduction objective

In our household, we can still improve several things. Because of this we will set a new objective at the beginning of each year. We will update our objective and put it into the SMART criteria format below.

SPECIFIC	Write a defined goal.
MEASURABLE	Can the outcome be measured?
ASSIGNABLE	Who will do it? Collaborate and get agreement with others.
REALISTIC	Is it achievable given available resources?
TIME FRAME	When will it be started and completed?

If you have completed the carbon footprinting process, you can choose to reduce your emissions by a specific amount or a percentage, for example:

- Reduce emissions by 2,000 metric tons from last year.
- Reduce emissions by 25% from last year.

SPECIFIC	Reduce emissions by an additional [new target] % from the level of last year. Then offset the remaining emissions to achieve net-zero.
MEASURABLE	Calculate emissions at the end to review progress.
ASSIGNABLE	We will work on reducing emissions together.
REALISTIC	The objective and time frame are both feasible.
TIME FRAME	Over the next 12 months.

Update the action plan

The next thing to do is plan additional actions and future progress. We found that the benefit of calculating our carbon footprint is that we now have information on the remaining emissions to help us decide where to act. Some things to consider when you are updating the action plan:

- Look at your carbon footprint and target areas of highest emissions.
- You can also review the checklists to identify actions that require additional progress.
- Identify which actions have the best opportunity for further reductions.
- You can also select some simple actions to start things off again and get some easy wins.

Before you start implementing the actions again, consider reviewing the section “Tips to Help You Be Successful” at the beginning of Part 3.

CONCLUSION

I want to acknowledge that when we struggle, it is normal to sometimes stumble and fall. Several times when we tried to reduce emissions, it took longer than expected or didn't go as planned. Even though I wanted to be a good example and reduce our emissions significantly, we had trouble making progress in several areas. This made me feel like I was failing.

Then I thought about the reasons for writing this book. I want to make the world a better place because I believe that this is the right thing to do. There is also a strong need for me not to be a bystander. I feel it is unacceptable to have knowledge that something is wrong and to do nothing. It's like watching someone drown and not even calling for help. I know there are many existing solutions available, and we can turn the situation around. I also feel that we have a strong moral responsibility to keep the conditions on Earth safe for present and future generations. Connecting with this knowledge and these feelings helped me get started again.

The important thing is to restart and then keep going. I began with simple things, creating easy wins, and building on success. I talked to my father, and we approached emissions reductions by working together.

One of the main things I have learned is that while reducing emissions and encouraging others to change, it may take some time and things may not always work out on the first try. Sometimes, when you are working on a long-term objective, setbacks and obstacles can get in the way. If we never give up and keep going forward, then we can't fail.

As you progress, I wish you well. I hope you will reduce your own emissions, encourage those around you, and influence others to take action now. I hope that an additional result will be that you feel more positive and confident about your future and life on Earth.

I strongly believe that as more people take action every day, it will create an overwhelming tide of change, making the world a better place and protecting our beautiful home.

Best wishes,
Philip

Keep updated about actions at www.climate-action.org.

A Clean Energy Future

Source: Zhongguo on iStock

Supporting the Author

I would be very grateful if you could take a few minutes to write a short review of this book. By sharing your thoughts, it may also assist other people decide if it might be useful to them. You can use whatever platform you prefer.

To write a quick review on Amazon, login and either:

1. Go to the top and click "Returns and Orders", then scroll to find the book, and click on the button "Write a product review"

or

2. Search for the name of the book and scroll down and look on the left, past the "Customer Reviews" until you see "Review this Product" and click on the button "Write a product review".

Your review will be very helpful, and I am interested to hear what you think.

Thank you,
Philip

APPENDICES

Appendix A: Oil Companies Knew

Obstacles

For some industries that need to respond to the climate crisis, a significant change is necessary. Consider the self-interest of an energy company whose assets are coal-fired power stations or oil and gas infrastructure. These facilities cost hundreds of millions of dollars and are costly to transform into renewable energy or another type of business. It might seem a lot easier and more profitable to do nothing or even to actively resist change. The following are a few documented examples of oil companies' resisting climate action.

1982: Oil and gas companies knew

In the 1980s, scientists at the oil companies Exxon and Shell reported that carbon dioxide levels had increased since the Industrial Revolution and that this was mainly caused by "fossil fuel burning and deforestation."[223] They estimated that if emissions continued to increase, the temperature would rise by 2°C by 2050 and 3°C by 2080.[224] Shell scientists reported, "Such relatively fast and dramatic changes would impact on the human environment, future living standards and food supplies, and could have major social, economic and political consequences."[225]

1998: American Petroleum Institute's "Roadmap"

The American Petroleum Institute helped organize a group that included oil companies Exxon and Chevron to develop a plan to promote the "uncertainties in climate science" to the public, the media, and politicians.[226] Their plan stated, "Unless 'climate change' becomes a non-issue, meaning that the Kyoto proposal is defeated and there are no further initiatives to thwart the threat of climate change, there may be no moment when we can declare victory."[227]

1998: Disinformation

Exxon published a pamphlet titled, "Global Climate Change: Everyone's Debate" and sent it to politicians in the United States and around the world. It contradicted their own scientific reports by saying, "Nearly all CO2 emissions come from natural sources. Only a small amount comes from burning fossil fuels."[228] Exxon also placed advertisements, of which 81% expressed doubt about climate change.[229]

2001: Kyoto failure

The United States didn't ratify the Kyoto treaty and dropped out in 2001.[230] The treaty only produced a commitment by 37 countries to reduce emissions by 5%.[231] This marked 16 years of diplomacy drowned by propaganda and disinformation.

2015 to 2018: Buying political power

By this point, promoting uncertainty was no longer viable, so some fossil fuel companies switched to obstruction. Investigators at Influence Map found that in the three years following the Paris Agreement, ExxonMobil, Royal Dutch Shell, Chevron, BP, and Total had invested more than $1 billion on climate-related lobbying.[232] It was also found that during the 2017 through 2018 midterm elections in the United States, the fossil fuel industry spent $359 million on lobbying and donations to national politicians and parties.[233]

2021: Greenwashing

New York City filed a lawsuit asserting that ExxonMobil, BP, Royal Dutch Shell, and the American Petroleum Institute systematically misled consumers about the central role their products play in causing the climate crisis.[234]

Appendix B: Impacts and Crisis Escalation

Impacts of Climate Change

Displacement

Rising sea levels are displacing people all over the world. Communities in Bangladesh, Fiji, Kiribati, Papua New Guinea, Solomon Islands, Tuvalu, the United States, and many others have lost their homes to sea inundation.[235] Every year, more and more people will lose their homes, and based on current projections 360 million people will be threatened by annual flood events by 2100 in a +2°C scenario.[236]

Studies led by NASA scientists show that the average global temperature on Earth has increased by at least 1.1°C (1.9°F) since 1880.[237] As temperatures continue to rise, many people will be living in areas that will become too hot for humans. This will affect an estimated 1.5 billion people by 2070 in southern Africa, Central and South America, Southeast Asia, Australia, India, Mexico, and the Middle East.[238] What happens if tens or hundreds of millions of climate refugees go on the move?[239] Climate change is affecting people all over the globe now, and with each passing year, it becomes everyone's problem.

The map below shows the number of people per country living on land expected to be under sea level by 2100. This is assuming a rise in sea levels of 50-70cm (2°C temperature increase/not taking into account ice sheet instability.[240]

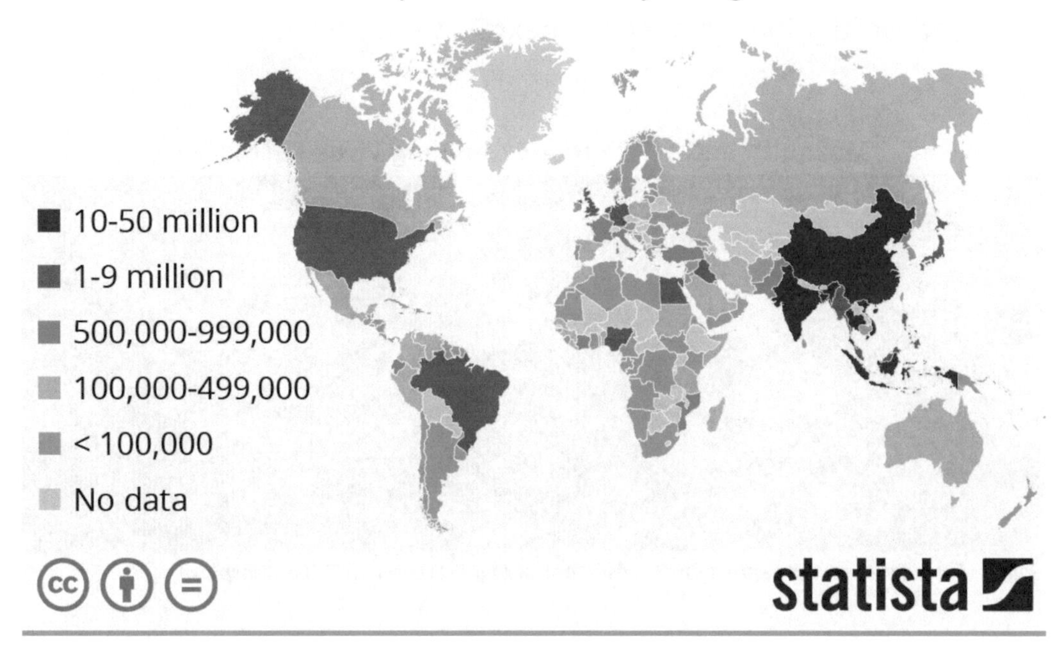

Source: Scott A. Kulp and Benjamin H. Strauss, Nature Communications [diagram by Statista]

Research by the United Nations showed that there is a vicious cycle, where people living in poverty suffer more than others from the adverse effects of climate change, resulting in even worse inequality.[241] People in poverty are less likely to afford insurance or be able to pay for repairs from the damage due to natural disasters.[242] These impacts are already pushing around 100 million additional people into poverty every year.[243] Therefore, climate action is also an issue of social justice.

Disease

During a pandemic, the last thing anyone wants to hear is that warming temperatures will mean more diseases. Yet researchers at Stanford University found that as it gets warmer, mosquitoes will roam beyond their current habitats.[244] This will increase the risk of dengue fever, zika and West Nile virus, for about 1 billion people in Europe, Russia, northern Asia, and North America over the next 50 years.[245]

Mosquito and a Scientist Examining Virus Particles

Source: Егор Камелев and Centers for Disease Control and Prevention (CDC) on Unsplash

Ecosystem destruction

Climate change is contributing to the collapse and death of entire ecosystems. Wildfires have been sweeping across the Amazon, Australia, Canada, China, Europe, and the United States, affecting wildlife and communities. In the polar regions, the loss of sea ice affects marine ecosystems, which includes polar bears, seals, birds, fish, and whales.[246]

Fire in the Amazon

Source: Brasil2 on iStock

Polar Bear Mother and Her Cub

Source: Alexey Seafarer on iStock

Ocean heatwaves are causing the mass die-off of coral reefs, kelp forests, seagrass, and mangroves.[247] Apart from processing large amounts of carbon dioxide, these ocean forests are the homes, shelter, and feeding grounds of thousands of marine species.[248] They act as nurseries for millions of fish that people rely on for food.[249] Living coral has declined by 50% and coral reef biodiversity has declined by at least 60% since the 1950s.[250] But all is not lost—there is still time to preserve and protect ecosystems and biodiversity.

A Healthy Coral Reef Ecosystem

Source: Olga Tsai on Unsplash

Dead Coral Reef from Bleaching

Source: Rich Carey on iStock

Is Climate Change an Emergency or a Crisis?

An emergency is "any incident, whether natural or human-caused, that requires responsive action to protect life or property."[251] People are losing their homes and lives now because of rising sea levels and worsening natural disasters.[252] Therefore, climate change should be considered an emergency.

A crisis, on the other hand, is a "situation that threatens an organization's strategic objectives, reputation, or viability."[253] Climate change threatens the way of life of communities now and will affect the entire human race in the coming decades, as well as increasing the threat to animals, plants, and the collapse of ecosystems. Considering this, climate change also represents a crisis for most life on our planet. We must recognize that if a crisis is managed properly and in time, then some impacts may be reduced, or avoided completely.

Crisis Escalation

Sea ice reflects up to 90% of solar energy.[254] However, rising temperatures cause sea ice to melt, and without ice cover, the sea absorbs most of the warmth from the sun.[255] This absorption then warms the ocean, which makes the ice melt faster, accelerating climate change.[256] The diagram on the right illustrates how melting sea ice leads to a positive feedback loop.[257]

Ice Melting Positive Feedback Loop

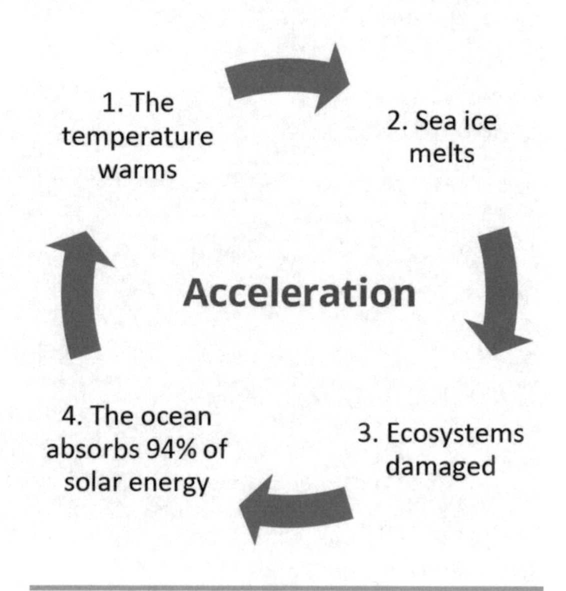

A review of satellite photos of the Earth has shown that between 1994 and 2017, more than 28,000 billion metric tons of ice has melted.[258] Ice loss is also a key driver of sea-level rise, which threatens hundreds of millions of people.[259]

Floating Ice in the Arctic Sea

Source: Rixipix on iStock

Permafrost is melting

In Alaska, Canada, northern Europe, and across Russia, ground that is frozen all year round, which is called permafrost, is melting. Locked in the frozen ground is an estimated 1,600 billion metric tons of carbon dioxide and methane.[260] These are gradually being released as the ground melts. This will speed up the warming, which will make the permafrost melt faster, which will release more CO_2 and methane. This is called a positive feedback loop, that is, the results increase the speed at which a system changes.[261]

Permafrost Melting Positive Feedback Loop

1. The temperature warms

2. Permafrost melts

Acceleration

3. Ecosystems and infrastruture damaged

4. CO_2 and methane released

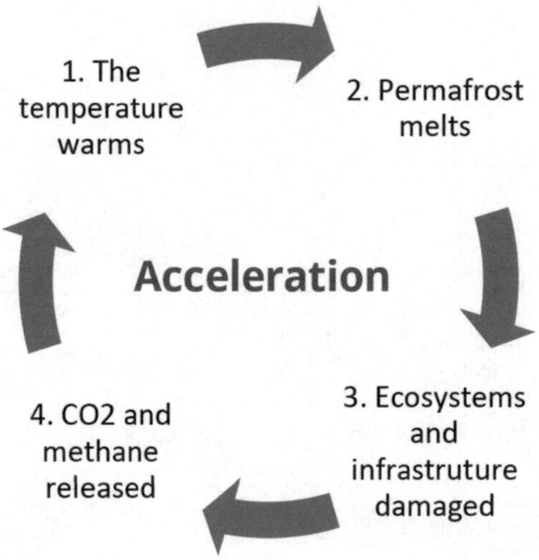

The IPCC estimates that 2 million square kilometers more permafrost will thaw this century with 2°C warming compared to 1.5°C—which is the same size as Mexico, making this another good reason to limit warming to 1.5°C.[262]

Drying and burning rainforests

A NASA study shows that "the atmosphere above the Amazon rainforest has been drying out, increasing the demand for water and leaving ecosystems vulnerable to fires and drought."[263]

As the temperature warms, rainforests become drier to the point where wildfires are happening, where previously it was not possible. More trees burn, die, and decompose, which releases carbon dioxide (CO_2). This then speeds up the warming, which creates a positive feedback loop, increasing system change.[264]

Drying Rainforests Positive Feedback Loop

1. The temperature warms

2. Rainforests become drier

Acceleration

3. Fires sweep through rainforests

4. CO_2 released and large trees die

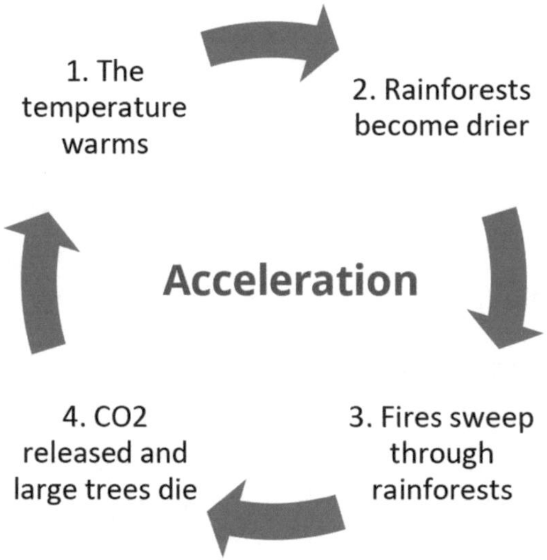

The NASA report concludes, "As the trees die, particularly the larger and older ones, they release CO_2 into the atmosphere; and the fewer trees there are, the less CO_2 the Amazon region would be able to absorb."[265]

Appendix C: Framework for Food

In each country, government at different levels will be responsible for policy and regulation of food production, distribution, retail, and service. The following outlines some basic elements that should be considered while respecting each region's differences, culture, and capacity.

Objectives, plan, and communication
The national, state, or local and city governments should

- Conduct regional and local community engagement to find the best solutions for policy, regulation, and support. This could be in-person, at town halls, or via online submission and should include farmers, indigenous people, transport, food processors, retailers, food service and consumers.
- Decide on an emissions reduction objective of 45% by 2030, then additional objectives for 2040 and 2050. Decide on a 50% reduction in food loss and waste by 2030, as this aligns with the SDG 12.3 objective.
- Develop a plan to meet the objectives, based on local and regional engagement. Communicate and promote the objectives and how they will be achieved through the plan to the community and agriculture industry.
- Provide policy stability so that households, farmers, and business can make a positive change in the future.

Support people and industry
The state and/or national governments should provide

- Subsidies or tax incentives for farmers to implement emissions reduction solutions.

- Public investment, loans, grants, capital subsidies for food transport, processing, or retail infrastructure to reduce food loss and waste.

Government policy

Ideally, government policy should:

- Provide funding for research and information programs to promote sustainable, low-carbon, healthy food lifestyles in the community.
- More than 80% of consumers in the United States occasionally discard food prematurely because of date label confusion with "best before," "display until," or "sell by."[266] Standardize food date labels so the "use-by" date is put most prominently on food so that consumers know about product safety. Also, provide consumer education.
- Legislate requirements on food retailers to donate food to charities instead of throwing it away. In 2016, France enacted legislation that required supermarkets to donate unwanted food to charities.[267] Within two years, 93% of retailers donated food to charity, compared to 33% before the law.[268] City and state governments should adjust regulations to make it easy for retailers to donate food. This should help to reduce hunger and improve access to safe and nutritious food, especially for poor and vulnerable people.

You can't fix what you don't understand

In 2021, the UN Environment Programme reported that only 17 countries had high-quality data about food loss or waste.[269] Not having enough information about the size, nature, and causes of a problem makes is difficult to find solutions. Governments in high-income countries should partner with developing countries in their region to help them measure food loss and waste. Then they should share the information with farmers, processors, retailers, consumers, and the international community. Each country should gather data using the United Nations Environmental Programme (UNEP) methodology.[270]

Support farmers
Create an integrated framework that supports farmers, the agriculture industry, researchers, retailers, and consumers to collaborate and collectively reduce emissions. The points on the following pages are some solutions that should be considered in a larger framework.

1. Implementation of existing solutions

The IPCC has identified solutions which are expected to have a medium or large impact on reducing emissions:[271]

- Improvements in cropland and livestock management and food productivity
- Reducing deforestation and expansion of agro forestry by planting trees and shrubs in crop and animal farming systems to create environmental, economic, and social benefits
- Increasing soil organic carbon content and adding biochar to soil to improve crop yields, water holding, and nutrient efficiency
- Reducing postharvest losses

2. Information gathering

Government should support the development of easy and streamlined ways that farmers can measure their carbon emissions. Several farming practices, from low tillage (digging or ploughing the soil) to agroforestry, can also increase carbon storage (sequestration) on the land. To encourage farmers to increase the level of carbon storage (sequestration) they also need easy and cost-effective ways to measure improvements.

This information on carbon emissions and storage should be shared with researchers.

3. Research into best practice

Promote research into low-emissions agriculture to help give farmers a range of options they can choose from. Contribute to an open global database to share information on best practice options for various crops in different regions and climates, in multiple languages. High-income countries should partner with developing countries in their region and assist with providing information and support to farmers to improve their crops and resilience while reducing emissions.

4. Support to developing countries

Food can be unnecessarily lost because farmers in developing countries don't have access to proper dry or cold storage for their crops. Storage solutions have been made at a low cost, and all that is needed is assistance from high-income countries to help distribute these. For example, Purdue University developed a system where grain is stored in interlocking plastic bags which keep out pests and keeps grain fresh for months.[272] These solutions will help small-scale producers to increase productivity and income, while reducing hunger and emissions.

5. Product feedback loop

Consumers should provide feedback on low- or zero-emissions food choices to retailers, processing companies, and food service. This will create demand and support farmers who are reducing their emissions.

6. Aiming for 100% utilization

There are a variety of different ways to use the parts of plants that are normally discarded during and after harvest.[273] This should be further researched to get as close to 100% utilization as possible and to share this knowledge. For example, the sugarcane stalk that's left after processing can be used in some packaging to replace plastic.[274] In addition, leftover organic material such as crop residues, manure, or weeds can be heated to 500°C to make bio-charcoal (also called biochar).[275] Biochar increases soil's ability to retain water, nutrients, and fertilizer, while also reducing the risk of soil erosion.[276] This means farmers can increase productivity while reducing fertilizers and emissions.[277]

7. Support farmers with the uptake of new technology

The government can support farms with grants, subsidies, or other types of tax credits to buy new technology, such as

- Solar panels to generate electricity and reduce running costs of the farm. Excess electricity can also be sold to the grid to improve annual farm income, especially in times of drought, floods, or fires.
- Zero-emissions electric tractors and other vehicles that can also help farmers save on fuel costs. Currently, these can manage jobs such as mowing and utility work that don't require heavy loads.[278] More research and development should be supported to make electric equipment that can handle more farming work.

We can employ many other approaches to support farmers and reduce emissions—this is just the beginning.

8. 50% less food loss objective by 2030

Farmers, transportation companies, food manufacturers, retailers, and consumers should work cooperatively to reduce food loss. For all participants, this could involve gathering information about food lost, then analyzing the reasons for the food loss and identifying opportunities for reductions.

Aerial View of a Solar Array on a Farm Growing Fruit and Vegetables

Source: baranozdemir on iStock

You can download this information in several formats from www.climate-action.org.

Appendix D: Framework for Industry

A great deal of information is required for a national or state industry emissions reduction plan. The following outlines elements that could be considered and adapted to local and regional circumstances.

Objectives, plan, and communication
The national or state governments should

- Conduct community and industry engagement to find the best solutions for policy, regulation, and support. This could be in-person, through town hall meetings, or via online submission.
- Decide on an emissions reduction objective of 45% by 2030 and additional objectives for 2040 and 2050. Then create an action plan and promote it widely.
- Provide policy stability so that business can plan its investments.

Replacing old with new
For several production processes, technology exists that can replace burning coal or gas, as you can see from the table on the next page.[279] Government should help industry to install low-emissions equipment and improve efficiency with financial incentives such as tax credits or deductions. Many of the zero-emissions technology also reduces production time and losses, requires lower maintenance, and has better safety.[280]

A Selection of Existing Zero-Emissions Solutions

HEATING TYPE	MATERIAL OR PRODUCT	EXISTING LOW-EMISSIONS SOLUTIONS
Drying	Electronics, textiles, paper, food, packaging, and automotive	Electric infrared
Cooking, sterilizing, and pasteurizing	Food and beverages	Heat pumps and electric infrared
Melting	Cement, glass, ceramics, and aluminum	Electric resistance furnace Electrical induction

Source: Beyond Zero Emissions

Support emerging solutions

Government and industry should invest in the development and widescale deployment of new solutions. Renewable hydrogen is an example of a new technology that is being developed to heat furnaces for making zero emissions steel. The first commercial steel production test projects started in 2020.[281] Fast-tracking this technology is important because world iron and steel production is responsible for 7% of total global emissions.[282]

Purchasing

Governments at all levels should purchase many types of low- and zero-emissions products to support the transition to a low-emissions economy.

Research and development
A collaboration should form between government, universities, and industry for research and commercial development of renewable industrial and manufacturing processes. The aim should be to take new technology to widescale implementation as quickly as possible. This should be shared across the world through licensing to developed or high-income countries and at no cost to developing countries.

Information sharing hub
Set up an information hub to make it easy for industry and manufacturing companies to find the best renewable solutions for their situations.

Regulatory framework
If industries or corporations are unwilling to change, then governments should gradually phase in a carbon tax for materials such as iron, steel, concrete, glass, and aluminum. This could start at a low level in 2025 and then gradually increase.

You can download this information in several formats from ***www.climate-action.org***.

Appendix E: Framework for Energy

A great deal of available information should go into a state or national energy plan. The following outlines some important elements that should be considered.

Objectives, plan, and communication

- Conduct community and industry engagement.
- Decide on renewable energy objectives for 2030, 2040 and 2050. Then develop a detailed plan to achieve the objectives and communicate and promote it to the public.
- As part of a just transition, the plan should consider the interests and rights of Indigenous people and those vulnerable to the impacts of climate change, as well as those who are affected by job loss as the industry adjusts.

Support people and industry

- All local, state, and national government buildings and facilities should make the switch to buying renewable energy or installing solar power. If governments in many countries did this, it would drive rapid change.
- Tax incentives, investment, or production tax credits for business investing in renewable energy.
- Public investment, loans, grants, and capital subsidies for grid upgrades, as well as rebates for consumers installing solar panels and batteries.

Promote development and energy efficiency

- Streamline regulations and grid connection to make it easy to develop renewable energy projects.
- Provide information to consumers and promote a choice of suppliers.
- Establish national energy efficiency standards, as well as promoting product energy efficiency and labeling.
- Technical deployment
- Conduct a survey to find the best areas for renewable energy such as solar, wind, and hydro that are near the electricity grid. Create maps and make them public.
- Identify the best locations for energy storage, such as pumped hydro and utility-scale battery installations.
- Upgrade the energy grid to promote power reliability, demand management, reduction of electricity losses, integrating renewable generation, and energy storage.
- Upgrade the energy grid to manage widely distributed solar and batteries being installed in households and businesses.
- Support a digitalized system including the deployment of smart metering. This helps consumers to become active participants who can use off-peak energy, sell, and optimize energy, and save money. This will also help manage demand response and grid stabilization.

Energy regulatory framework

- Set a standard requiring that a minimum percentage of generation sold by a utility is provided by renewable energy. It could be set low, for example, 5%, and then given a reasonable time frame of five years to gradually increase. This is often called a Renewable Portfolio Standard (RPS).
- Provide a feed-in tariff: minimum payments to households, communities, and businesses for generating electricity and supplying it into the grid.

Support developing countries

High-income countries should partner with developing countries in their region and assist with providing technical support and loans to improve access of their people to affordable zero-emissions electricity.

You can download this information in several formats from www.climate-action.org.

Appendix F: Framework for Transportation

Each country shares responsibilities for transportation between national, state, or local government levels in different ways. The following outlines some elements that should be considered.

Objectives, plan, and communication
The national, state, or local and city governments should

- Conduct regional and local community engagement to find the best solutions for infrastructure, regulation, and support.
- Decide on renewable energy objectives for 2030, 2040, and 2050.
- Develop a plan.
- Communicate and promote to the nation the objectives and the plan and how they will be achieved.
- Provide policy stability so that households and business can plan investments in renewable energy.

Make long-term planning normal
It should be against the law for a state or national government not to have a long-term integrated transportation plan that looks forward at least ten years. Election cycles, where leaders are only in power between three and four years, encourage leaders to focus on getting a handful of projects completed before they need to think about the lead-up to the next election. The consequence is that long-term planning (+10 to 20 years) and ongoing planning for cities and towns that are experiencing rapid expansion is often not done. One option could be that an ongoing committee of political representatives from the major parties and independents should be formed. The intention would be to create long-term plans for public transportation and infrastructure.

Promote and expand public transport

More people are expected to live in urban areas, increasing from 55% in 2020 to 68% in 2050.[283] Cities can use this as an opportunity to expand their public transportation, which would reduce congestion on roads. In highly populated areas, mass transit is the most efficient option, saving time and significant costs to the economy.[284] Decreases in travel time will also reduce costs, fuel or energy, and wear on road vehicles. Governments should invest in mass transit and encourage private investment with tax incentives or subsidies.

Switch public transport

Government could also mandate that all public transportation that can run on electricity, such as trains and streetcars, should be powered by renewable energy. State-owned public transportation should be switched to zero emissions. Privately owned public transportation should be encouraged to switch with subsidies, tax incentives, and, if necessary, regulation. Phase in electrified or hydrogen-powered buses.

Incentivize zero-emissions vehicle ownership

Currently, some zero-emissions vehicles have a higher upfront cost compared to older internal combustion engine vehicles. The costs of zero-emissions vehicles will decrease faster if the industry is supported. Governments could create purchase incentives by reducing or removing registration fees, import tariffs, sales tax, stamp duty, or other fees on zero-emissions vehicles. This revenue could be recovered by applying additional fees and taxes on luxury cars with internal combustion engines.

Government leadership

Local, state, and national governments should increase the number of zero-emissions vehicles in their fleets. Governments should set a policy where all new government vehicles must be zero emissions. If governments in many countries did this, it would drive the rapid change. This should include setting up a charging infrastructure at government locations.

Phase out fossil fuel combustion engines

Many countries and cities around the world have proposed dates for phasing out the sale of passenger vehicles powered by fossil fuels. This could take the form of gradually increasing sales taxes, or prohibition on the importation, production, or sales of new vehicles with internal combustion engines. These could be phased in; for example, all car and motorcycle sales by 2030, and all bus and truck sales by 2035.

Building standards to promote zero-emissions infrastructure

Governments should update the relevant building and planning requirements to ensure that all new commercial developments are equipped with electric vehicle charging. Supermarkets and shopping centers should be encouraged to install recharge points. Interstate highways and regional towns should also have electric vehicle charging infrastructure to support long-distance driving.

Promote active transport

Active transportation is human powered, such as walking, cycling, e-bikes, and e-scooters. Government should provide adequate infrastructure to help people travel safely. This can include crosswalks, overpasses, sidewalks, bike paths, and bike lanes on roads. State and local laws can also contribute to protecting pedestrians and cyclists and improve the safety of active transportation.

Drive innovation now

The government should signal to rail, air, and sea transportation industries that they must do more now to speed up the development of zero-emissions solutions. National and state governments can support innovation with subsidies, investment, and tax reduction incentives.

You can download this information in several formats from ***www.climate-action.org.***

ENDNOTES

1. Secretary-General António Guterres. "Secretary-General's Remarks on ClimateChange [as delivered]." 10 September 2018, United Nations, https://www.un.org/sg/en/content/sg/statement/2018-09-10/secretary-generals-remarks-climate-changedelivered. Accessed 21 September 2019.

2. United Nations General Assembly. "Protection of Global Climate for Present and Future Generations of Mankind (A/45/696/Add.1)." 8 November 1990, https://unfccc.int/sites/default/files/resource/docs/1990/un/eng/a45696a1.pdf. Accessed 24 April 2021.

3. American Petroleum Institute (API). "Draft Global Climate Science Communications Plan." 3 April 1998, http://www.climatefiles.com/trade-group/american-petroleuminstitute/1998-global-climate-science-communications-team-action-plan/. Accessed 22 July 2022.

4. Ibid

5. Influence Map. "How the Oil Majors Have Spent $1Bn Since Paris on Narrative Capture and Lobbying on Climate." March 2019, InfluenceMap, https://influencemap.org/report/How-Big-Oil-Continues-to-Oppose-the-Paris-Agreement-38212275958aa21196dae3b76220bddc. Accessed 7 April 2021.

6. IPCC. "The Evidence is Clear: The Time for Action is Now. We Can Halve Emissions by 2030." 4 April 2022, IPCC Press Release, https://report.ipcc.ch/ar6wg3/pdf/IPCC_AR6_WGIII_PressRelease-English.pdf. Accessed 5 June 2022.

7. Nicholas Stern, *Stern Review: The Economics of Climate Change*, Cabinet Office, HM Treasury. https://webarchive.nationalarchives.gov.uk/ukgwa/20100407172811/https:/www.hm-treasury.gov.uk/stern_review_report.htm.

8. Nicholas Stern, "Climate Change, ethics and the economics of the global deal," *VoxEU-CEPR*, 30 November 2007. https://cepr.org/voxeu/columns/climate-change-ethics-and-economics-global-deal.

9. Rebecca Doherty, Claudia Kampel, Anna Koivuniemi, Lucy Pérez, and Werner Rehm, "The triple play: Growth, profit, and sustainability," *McKinsey and Company*, 9 August 2023. https://www.mckinsey.com/capabilities/strategy-and-corporate-finance/our-insights/the-triple-play-growth-profit-and-sustainability.

10. John Vidal, Suzanne Goldenberg, and Lenore Taylor. "How the historic Paris deal over climate change was finally agreed," *The Guardian*, 13 December 2015. https://www.theguardian.com/environment/2015/dec/13/climate-change-deal-agreed-paris.

11. A. Leiserowitz, E. Maibach, S. Rosenthal, J. Kotcher, E. Goddard, J. Carman, M. Verner, M. Ballew, J. Marlon, S. Lee, T. Myers, M. Goldberg, N. Badullovich, K. Their, "Climate

Change in the American Mind: Politics & Policy," Yale University and George Mason University, Fall 2023, New Haven, CT: Yale Program on Climate Change Communication. https://climatecommunication.yale.edu/publications/climate-change-in-the-american-mind-politics-policy-fall-2023/.

12. Ibid

13. Ibid.

14. Brayden G. King (2008), "A political mediation model of corporate response to social movement activism," *Administrative Science Quarterly, 53*(3), 395-421. https://doi.org/10.2189/asqu.53.3.395.

15. Connie Roser-Renouf, Edward Maibach and Anthony Leiserowitz,"The Consumer as Climate Activist," *International Journal of Communication*, 14 October 2016, 10, https://climatecommunication.yale.edu/publications/consumer-activism-global-warming/.

16. Climate Action Tracker (2023), *The CAT Thermometer*, December 2023. https://climateactiontracker.org/global/cat-thermometer/. Accessed 11 May 2024.

17. Robert McSweeney, "Analysis: The gender, nationality and institution of IPCC AR6 scientists, "15 May 2018, *Carbon Brief*. https://www.carbonbrief.org/analysis-gender-nationality-institution-ipcc-ar6-authors/.

18. IPCC, *Summary for Policymakers of IPCC Special Report on Global Warming of 1.5°C approved by governments*, 8 October 2018, IPCC website. https://www.ipcc.ch/2018/10/08/summary-for-policymakers-of-ipcc-special-report-on-global-warming-of-1-5c-approved-by-governments/

19.United Nations, *Climate Change-induced Sea-Level Rise Direct Threat to Millions around World, Secretary-General Tells Security Council*, 14 February 2023, SC/15199. https://press.un.org/en/2023/sc15199.doc.htm

20. Ibid

21. Kulp, S.A., and B.H. Strauss. "New Elevation Data Triple Estimates of Global Vulnerability to Sea-Level Rise and Coastal Flooding." Nature Communication, vol. 10, 29 October 2019, https://doi.org/10.1038/s41467-019-12808-z. Accessed 6 March 2021. Helen Horton, "Mosquito-borne diseases spreading in Europe due to climate crisis, says expert," *The Guardian*, 25 April 2024. https://www.theguardian.com/environment/2024/apr/25/mosquito-borne-diseases-spreading-in-europe-due-to-climate-crisis-says-expert.

22. Lustgarten, Abrahm. "The Great Climate Migration." 23 July 2020, The New York Times, https://www.nytimes.com/interactive/2020/07/23/magazine/climate-migration.html. Accessed 15 March 2020.

23. Ryan, Sadie J., Colin J. Carlson, Erin A. Mordecai, and Leah R. Johnson. "Global Expansion and Redistribution of Aedes-Borne Virus Transmission Risk with Climate Change." PLOS Neglected Tropical Diseases, 28 March 2019, https://journals.plos.org/plosntds/article?id=10.1371/journal.pntd.0007213.

24. Jordan, Rob. "How Does Climate Change Affect Disease?" March 15, 2019, Stanford Woods Institute for the Environment. https://earth.stanford.edu/news/how-doesclimate-change-affect-disease. Accessed 13 March 2021; Ryan, Sadie J., et al. "Global Expansion and Redistribution of Aedes-Borne Virus Transmission Risk with Climate Change." PLOS Neglected Tropical Diseases, 28 March 2019, https://journals.plos.org/plosntds/article?id=10.1371/journal.pntd.0007213. Accessed 2 January 2022.

25. Pagano, Anthony M., and Terrie M. Williams, "Physiological Consequences of Arctic Sea Ice Loss on Large Marine Carnivores: Unique Responses by Polar Bears and Narwhals." The Journal of Experimental Biology, vol. 224 (Suppl_1), February 2021, https://journals.biologists.com/jeb/article/224/Suppl_1/jeb228049/237178/Physiological-consequencesof-Arctic-sea-ice-loss. Accessed 4 January 2022.

26. Layton, Cayne, et al. "Kelp Forest Restoration in Australia." 14 February 2020, Frontiers in Marine Science, https://www.frontiersin.org/articles/10.3389/fmars.2020.00074/full. Accessed 17 April 2021; Australian Government Great Barrier Reef Marine Park Authority. "Climate Change." https://www.gbrmpa.gov.au/our-work/threats-to-the-reef/climate-change.

27. IPCC, *Summary for Policymakers of IPCC Special Report on Global Warming of 1.5°C approved by governments*, 8 October 2018, IPCC website. https://www.ipcc.ch/2018/10/08/summary-for-policymakers-of-ipcc-special-report-on-global-warming-of-1-5c-approved-by-governments/

28. Dr. Jane Goodall, Jane Goodall Institute USA Official Store, https://shop.janegoodall.org/product/every-individual-makes-a-difference-quote-pin/jgi277

29. IPCC. "Special Report: Global Warming of 1.5 °C: Summary for Policymakers." 2018, https://www.ipcc.ch/sr15/chapter/spm/. Accessed 3 January 2022.

30. Damon Centola, Joshua Becker, Devon Brackbill, and Andrea Baronchelli, "Experimental evidence for tipping points in social convention," 8 June 2018, *Science, 360* (6393), 1116-1119. https://www.science.org/doi/10.1126/science.aas8827. Accessed 2 June 2024.

31. Carrington, Damian. "UN Global Climate Poll: 'The People's Voice is Clear – They Want Action.'" The Guardian, 27 January 2021, https://www.theguardian.com/environment/2021/jan/27/un-global-climate-poll-peoples-voice-is-clear-they-wantaction. Accessed 15 August 2021.

32. Craw, Victoria. "Climate Strike is Best Excuse Yet to Stay Home Today." News, 30 November 2015, https://www.news.com.au/technology/environment/climate-change/climate-strike-is-best-excuse-yet-to-stay-home-today/news-story/e24b93b9decd79347d21f958d75e74f8. Accessed 21 May 2022.

33. Haynes, Suyin. "Students From 1,600 Cities Just Walked Out of School to Protest Climate Change. It Could Be Greta Thunberg's Biggest Strike Yet." Time, 24 May 2019, https://time.com/5595365/global-climate-strikes-greta-thunberg/. Accessed 20 June 2021.

34. Steffen, Will, et al. "Planetary Boundaries: Guiding Human Development on a Changing Planet." Science, 13 February 2015, https://www.science.org/doi/10.1126/science.1259855. Accessed 15 November 2021; Stockholm Resilience Centre. "Planetary Boundaries." Stockholm University, January 2022, https://www.stockholmresilience.org/research/planetary-boundaries.html. Accessed 20 March 2022.

35. United Nations Sustainable Development. "Report of the World Commission on Environment and Development - Our Common Future." https://sustainabledevelopment.un.org/milestones/wced, Accessed 18 September 2021.

36.Ibid.

37. Ibid.

38.Ibid.

39.United Nations. "Sustainable Development Goals Officially Adopted by 193 Countries," 23 September 2015, http://www.un.org.cn/info/6/620.html. Accessed 5 July 2021.

40. IPCC. "Foreword." Special Report: Global Warming of 1.5 °C: Impacts of 1.5°C Global Warming on Natural and Human Systems, 2018, https://www.ipcc.ch/sr15/about/foreword/. Accessed 3 February 2022.

41.United Nations. "THE 17 GOALS." United Nations Department of Economic and Social Affairs Sustainable Development, https://sdgs.un.org/. Accessed 13 July 2021.

42. Whitmarsh, Lorraine. "Moments of Change for Pro-Environmental Behaviour Shifts (MOCHA)." Cardiff University - School of Psychology, 2019, https://www.cardiff.ac.uk/psychology/research/impact/moments-of-change-mocha. Accessed 16 November 2021.

43. Lally, Phillippa, Cornelia H. M. Van Jaarsveld, Henry W. W. Potts, and Jane Wardle. "How Are Habits Formed: Modelling Habit Formation in the Real World." European Journal of Social Psychology, 16 July 2009, http://repositorio.ispa.pt/bitstream/10400.12/3364/1/IJSP_998-1009.pdf. Accessed 15 August 2021.

44. Duckworth, Angela. Grit: The Power of Passion and Perseverance. Vermilion – Mass Market, 1 May 2017.

45. Dweck, Carol. *Mindset - Updated Edition: Changing the Way You Think to Fulfil Your Potential*. Little, Brown Book Group, 12 January 2017, p. 250.

46.Ibid.

47. Dweck, Carol. "The Power of Believing that You Can Improve," TEDx Norrkoping, www.ted.com/talks/carol_dweck_the_power_of_believing_that_you_can_improve. Accessed July 15, 2020.

48. IPCC. "IPCC Special Report on Climate Change, Desertification, Land Degradation, Sustainable Land Management, Food Security, and Greenhouse gas fluxes in Terrestrial Ecosystems - Summary for Policymakers." 2018, IPCC, https://www.ipcc.ch/site/assets/uploads/2019/08/3.-Summary-of-Headline-Statements.pdf. Accessed 25 August 2019.

49. United Nations. "End Hunger, Achieve Food Security and Improved Nutrition and Promote Sustainable Agriculture." United Nations Department of Economic and Social Affairs Statistics Division, 2020, https://unstats.un.org/sdgs/report/2021/goal-02/. Accessed 12 February 2022.

50. Hannah Ritchie, "Emissions by sector", Our World in Data, https://ourworldindata.org/ghg-emissions-by-sector, Accessed 29 May 2024; Stein Emil Vollset, Emily Goren, Chun-Wei Yuan, Jackie Cao, Amanda E Smith, Thomas Hsiao, et al. "Fertility, mortality, migration, and population scenarios for 195 countries and territories from 2017 to 2100: a forecasting analysis for the Global Burden of Disease Study." The Lancet. 14 July 2020. https://www.thelancet.com/article/S0140-6736(20)30677-2/fulltext. Accessed 11 August 2021; Intergovernmental Panel on Climate Change (IPCC). "Special Report: Special Report on Climate Change and Land – Food Security." January 2020. https://www.ipcc.ch/srccl/chapter/chapter-5/. Accessed 13 August 2021.

51. United Nations. "'Foodkit': Our Global Anti-Food-Waste Campaign Toolkit." May 2018, United Nations Environment Programme, https://wedocs.unep.org/bitstream/handle/20.500.11822/25183/TES_FoodKit ToolKit_WEB.pdf. Accessed 29 August 2021; United Nations. "UNEP Food Waste Index Report 2021." United Nations Environment Programme, 4 March 2021, www.unep.org/resources/report/unep-food-waste-indexreport-2021. Accessed 8 August 2021.

52. United Nations. "'Foodkit': Our Global Anti-Food-Waste Campaign Toolkit." United Nations Environment Programme, 4 March 2021, www.unep.org/resources/report/unep-food-waste-index-report-2021. Accessed 8 August 2021.

53. United Nations. "Goal 12: Ensure Sustainable Consumption and Production Patterns." United Nations Sustainable Development, 2019, https://www.un.org/sustainabledevelopment/sustainable-consumption-production/. Accessed 25 August 2021; United Nations. "'Foodkit': Our Global Anti-Food-Waste Campaign Toolkit." United

Nations Environment Programme, 4 March 2021, www.unep.org/resources/report/unepfood-waste-index-report-2021. Accessed 8 August 2021.

54. United Nations. "Goal 12: Ensure Sustainable Consumption and Production Patterns." United Nations Sustainable Development, 2019, https://www.un.org/sustainabledevelopment/sustainable-consumption-production/. Accessed 25 August 2021.

55. Foley, Jonathan, Jim Richardson, and George Steinmetz. "Where Will We Find Enough Food to Feed 9 Billion?" National Geographic, 2014, https://www.nationalgeographic.com/foodfeatures/feeding-9-billion/. Accessed 28 August 2021.

56. Bajželj, Bojana, Keith S. Richards, Julian M. Allwood, Pete Smith, John S. Dennis, Elizabeth Curmi, and Christopher A. Gilligan. "Importance of Food-Demand Management for Climate Mitigation." Nature, 31 August 2014, https://www.nature.com/articles/nclimate2353. Accessed 11 August 2021.

57. Mrówczyńska-Kamińska, Aldona, Bartłomiej Bajan, Krzysztof Piotr Pawłowski, Natalia Genstwa and Jagoda Zmyślona. "Greenhouse Gas Emissions Intensity of Food Production Systems and Its Determinants." PLOS, 30 April 2021, https://journals.plos.org/plosone/article?id=10.1371/journal.pone.0250995. Accessed 13 August 2021; IPCC. "Food Security." Special Report: Special Report on Climate Change and Land, The Intergovernmental Panel on Climate Change, January 2020, https://www.ipcc.ch/srccl/chapter/chapter-5/. Accessed 13 August 2021.

58. Food and Agriculture Organization of the United Nations (FAO). "The Future of Food and Agriculture – Alternative Pathways to 2050." 2018, https://www.fao.org/3/CA1553EN/ca1553en.pdf. Accessed 2 February 2022.

59. IPCC, "Food Security." Special Report: Special Report on Climate Change and Land, The Intergovernmental Panel on Climate Change, January 2020, https://www.ipcc.ch/srccl/chapter/chapter-5/. Accessed 13 August 2021.

60. IPCC. "Figure 5.12." Special Report on Climate Change and Land Demand Side Mitigation - Greenhouse Gas Mitigation Potential of Different Diets, Intergovernmental Panel on Climate Change, 8 August 2019, https://www.ipcc.ch/srccl/chapter/chapter-5/5-5-mitigationoptions-challenges-and-opportunities/5-5-2-demand-side-mitigation-options/5-5-2-1-mitigation-potential-of-different-diets/figure-5-12/. Accessed 4 November 2021.

61. IPCC. "Chapter 3." Special Report: Global Warming of 1.5 °C: Impacts of 1.5°C Global Warming on Natural and Human Systems, Intergovernmental Panel on Climate Change, 2018, https://www.ipcc.ch/sr15/chapter/chapter-3/. Accessed 12 May 2021.

62. Department of Economic and Social Affairs: Indigenous Peoples. "United Nations Declaration on the Rights of Indigenous Peoples," 2 October 2007, https://www.un.org/

development/desa/indigenouspeoples/declaration-on-the-rights-of-indigenouspeoples.html. Accessed: 11 January 2022.

63. Consortium of International Agricultural Research Centers (CGIAR) and Springmann, Marco, et al., "Options for Keeping the Food System within Environmental Limits." Nature, 10 October 2018, https://www.nature.com/articles/s41586-018-0594-0.epdf. Accessed 14 August 2021.

64. Poore, J., and T. Nemecek. "Reducing Food's Environmental Impacts through Producers and Consumers." Science, 1 June 2018, https://science.sciencemag.org/content/360/6392/987. Accessed 15 August 2021.

65. Ritchie, Hannah. "You Want to Reduce the Carbon Footprint of Your Food? Focus on What You Eat, Not Whether Your Food is Local." Our World in Data, 24 January 2020, https://ourworldindata.org/food-choice-vs-eating-local; Poore, J., and T. Nemecek. "Reducing Food's Environmental Impacts through Producers and Consumers." Science, 1 June 2018, https://science.sciencemag.org/content/360/6392/987. Accessed 15 August 2021.

66. Ritchie, Hannah. "The Carbon Footprint of Foods: Are Differences Explained by the Impacts of Methane?" 10 March 2020, Our World in Data, https://ourworldindata.org/carbon-footprint-food-methane, Accessed 19 June 2020.

67. National Chicken Council. "Per Capita Consumption of Poultry and Livestock, 1965 to Forecast 2022, in Pounds." The National Chicken Council (United States), September 2021, https://www.nationalchickencouncil.org/about-the-industry/statistics/per-capita consumption-of-poultry-and-livestock-1965-to-estimated-2012-in-pounds/. Accessed 2 November 2021.

68. Ritchie, Hannah. "You Want to Reduce the Carbon Footprint of Your Food? Focus on What You Eat, Not Whether Your Food is Local." Our World in Data, 24 January 2020, https://ourworldindata.org/food-choice-vs-eating-local. Accessed 29 March 2020; Poore, J., and T. Nemecek. "Reducing Food's Environmental Impacts through Producers and Consumers." Science, 1 June 2018, https://science.sciencemag.org/content/360/6392/987. Accessed 15 August 2021.

69. IPCC. "Food Security." Special Report: Special Report on Climate Change and Land, Intergovernmental Panel on Climate Change, January 2020, https://www.ipcc.ch/srccl/chapter/chapter-5/. Accessed 13 August 2021.

70. IPCC. "Figure 5.12." Special Report on Climate Change and Land Demand Side Mitigation - Greenhouse Gas Mitigation Potential of Different Diets, Intergovernmental Panel on Climate Change, 8 August 2019, https://www.ipcc.ch/srccl/chapter/chapter-5/5-5-mitigationoptions-challenges-and-opportunities/5-5-2-demand-side-mitigation-

options/5-5-2-1-mitigation-potential-of-different-diets/figure-5-12/. Accessed 4 November 2021.

71. Ibid.

72. U.S. Department of Agriculture and U.S. Department of Health and Human Services. "Dietary Guidelines for Americans 2020–2025." December 2020, https://www.dietaryguidelines.gov/sites/default/files/2020-12/Dietary_Guidelines_for_Americans_2020-2025.pdf. Accessed 5 November 2021.

73. Ibid.

74. U.S. Department of Agriculture and U.S. Department of Health and Human Services. "What is MyPlate?" December 2020, https://www.myplate.gov/eat-healthy/what-ismyplate. Accessed 5 November 2021.

75. United States Department of Agriculture. "MyPlate: Style Guide and Conditions of Use for the Icon." 2020, https://myplate-prod.azureedge.net/sites/default/files/2021-01/MyPlateStyleGuide_2020-2025.pdf. Accessed 22 January 2022.

76. Ritchie, Hannah. "You Want to Reduce the Carbon Footprint of Your Food? Focus on What You Eat, Not Whether Your Food is Local." Our World in Data, 24 January 2020, https://ourworldindata.org/food-choice-vs-eating-local. Accessed 29 March 2020.

77. UN Environment Programme. "Food Waste Index Report 2021." 2021, https://www.unep.org/resources/report/unep-food-waste-index-report-2021. Accessed 6 November 2021; UN Environment Programme. "Worldwide Food Waste." https://www.unep.org/thinkeatsave/get-informed/worldwide-food-waste. Accessed 6 November 2021.

78. Ibid

79. UN Environment Programme. "Definition of Food Loss and Waste." https://www.unep.org/thinkeatsave/about/definition-food-loss-and-waste. Accessed 9 August 2021.

80. Ibid.

81. UN Environment Programme. "UNEP Food Waste Index Report 2021." https://www.unep.org/resources/report/unep-food-waste-index-report-2021, p. 70. Accessed 6 November 2021. [Please note that the figure of 369 million for the supply side is derived from taking 1.3 billion and then subtracting 931 million of food wasted.]

82. United Nations Department of Economic and Social Affairs. "12 Ensure Sustainable Consumption and Production Patterns." https://sdgs.un.org/goals/goal12. Accessed 6 November 2021.

83. Food Safety and Inspection Service (USDA). "Food Product Dating." 2 October 2019, https://www.fsis.usda.gov/food-safety/safe-food-handling-and-preparation/foodsafety-basics/food-product-dating. Accessed 19 June 2022.

84. Zeratsky, Katherine. "How Long Can You Safely Keep Leftovers in the Refrigerator?" Mayo Clinic, September 2020, p. 29, https://www.mayoclinic.org/healthy-lifestyle/nutrition-and-healthy-eating/expert-answers/food-safety/faq-20058500. Accessed 31 August 2021.

85.Thompson, Richard. "Gardening for Health: A Regular Dose of Gardening." Clinical Medicine, vol. 18, no. 3, June 2018, pp. 201-205, https://www.ncbi.nlm.nih.gov/pmc/articles/PMC6334070/. Accessed July 21, 2022.

86. United Nations Food and Agriculture Organization. "Small Family Farmers Produce a Third of the World's Food." 23 April 2021, https://www.fao.org/news/story/en/item/1395127/icode/. Accessed 21 November 2021.

87. Poore, J., and T. Nemecek. "Reducing Food's Environmental Impacts through Producers and Consumers." Science, 1 June 2018, https://science.sciencemag.org/content/360/6392/987. Accessed 15 August 2021.

88. Ibid.

89. Ibid.

90. Mooney, Pat. Too Big to Feed: Exploring the Impacts of Mega-Mergers, Concentration, Concentration of Power in the Agri-Food Sector. IPES-Food, October 2017, https://www.ipes-food.org/_img/upload/files/Concentration_FullReport.pdf. Accessed 22 November 2021; Kor, Yasemin Y., Jaideep Prabhu, and Mark Esposito, "How Large Food Retailers Can Help Solve the Food Waste Crisis," Harvard Business Review, 19 December 2017, https://hbr.org/2017/12/how-large-food-retailers-can-help-solve-the-food-waste-crisis. Accessed 22 November 2021.

91. UN Environment Programme. "UNEP Food Waste Index Report 2021." https://www.unep.org/resources/report/unep-food-waste-index-report-2021, p. 70. Accessed 6 November 2021. [Please note that the figure of 369 million for the supply side is derived from taking 1.3 billion and then subtracting 931 million of food wasted.]

92. Kenton, Will. "Social License to Operate (SLO)?" Investopedia, 31 May 2021, https://www.investopedia.com/terms/s/social-license-slo.asp. Accessed 27 November 2021.

93.Napoletano, E., and Benjamin Curry. "Environmental, Social and Governance: What Is ESG Investing?" Forbes, https://www.forbes.com/advisor/investing/esg-investing/. Accessed 7 February 2022.

94. Ibid.

95. OECD. "Global Material Resources Outlook to 2060." 12 February 2019. https://www.oecd.org/env/global-material-resources-outlook-to-2060-9789264307452-en.htm. Accessed 10 September 2021; Hannah Ritchie and Max Roser. "Emissions by

sector." Our World in Data. https://ourworldindata.org/emissions-by-sector. Accessed 18 February 2021; OECD. "Global Material Resources Outlook to 2060." 12 February 2019. https://read.oecd-ilibrary.org/environment/global-material-resources-outlook-to-2060_9789264307452-en#page122. Accessed 10 September 2021

96. Steffen, Will, Wendy Broadgate, Lisa Deutsch, Owen Gaffney, and Cornelia Ludwig. "The Trajectory of the Anthropocene: The Great Acceleration." The Anthropocene Review, 16 January 2015, https://doi.org/10.1177/2053019614564785. Accessed 4 January 2022.

97. Motesharrei, Safa, et al. "Modeling Sustainability: Population, Inequality, Consumption, and Bidirectional Coupling of the Earth and Human Systems." National Science Review, December 2016, https://doi.org/10.1093/nsr/nww081. Accessed 4 January 2022.

98. Earth Overshoot Day. "About Earth Overshoot Day." 2021, https://overshoot.footprintnetwork.org/. Accessed 4 January 2024.

99. Ibid.

100. Global Footprint Network. "Ecological Footprint." 2021, https://www.footprint network.org/our-work/ecological-footprint/. Accessed 4 January 2022.

101. OECD. "Global Material Resources Outlook to 2060." OECD, 12 February 2019, p.122, https://read.oecd-ilibrary.org/environment/global-material-resources-outlook-to-2060_9789264307452-en#page122. Accessed 10 September 2021.

102. Ritchie, Hannah, and Max Roser. "Emissions by Sector." Our World in Data, https://ourworldindata.org/emissions-by-sector. Accessed 18 February 2021; Olivier, J.G.J., and J.A.H.W. Peters. "Trends in Global CO2 and Total Greenhouse Gas Emissions 2019." PBL Netherlands Environmental Assessment Agency, May 2020, https://www.pbl.nl/sites/default/files/downloads/pbl-2020-trends-in-global-co2-and-total-greenhouse-gasemissions-2019-report_4068.pdf. Accessed 13 June 2021.

103. United Nations Industrial Development Organization. "Driving Towards Circularity." November 2017, https://www.unido.org/sites/default/files/files/2018-05/20180503Conference report.pdf. Accessed: 14 August 2022.

104. United Nations Industrial Development Organization (UNIDO). "Circular Economy." 2017, https://www.unido.org/sites/default/files/2017-07/Circular_Economy_UNIDO_0.pdf. Accessed 11 September 2021.

105. World Bank. "WHAT A WASTE 2.0 – A Global Snapshot of Solid Waste Management to 2050." 20 September 2018, https://datatopics.worldbank.org/what-a-waste/trends_in_solid_waste_management.html. Accessed 22 August 2021.

106. Yorke, Liselle. "2021 World Population Data Sheet Release." Population Reference Bureau (PRB), 17 August 2021, https://www.prb.org/news/2021-world-population-datasheet-released/. Accessed 7 February 2022.

107. United Nations Industrial Development Organization (UNIDO). "Circular Economy." 2017, https://www.unido.org/sites/default/files/2017-07/Circular_Economy_UNIDO_0.pdf. Accessed 11 September 2021.

108. United Nations Industrial Development Organization. "Circular Economy." 2022, https://www.unido.org/sites/default/files/2017-07/Circular_Economy_UNIDO_0.pdf. Accessed: 14 August 2022.

109. IPCC. "Chapter 3." Special Report: Global Warming of 1.5 °C: Impacts of 1.5°C Global Warming on Natural and Human Systems, IPCC, 2018, https://www.ipcc.ch/sr15/chapter/chapter-3/. Accessed 12 May 2021.

110. United Nations Industrial Development Organization (UNIDO). "Circular Economy." 2017, https://www.unido.org/sites/default/files/2017-07/Circular_Economy_UNIDO_0.pdf. Accessed 11 September 2021.

111. Farrell, Sean. "We've Hit Peak Home Furnishings, Says Ikea Boss." The Guardian, 18 Jan 2016, https://www.theguardian.com/business/2016/jan/18/weve-hit-peak-homefurnishings-says-ikea-boss-consumerism. Accessed 12 September 2021.

112. Deloitte Insight. "The Growing Power of Consumers." Deloitte, 2014, https://www2.deloitte.com/content/dam/Deloitte/uk/Documents/consumer-business/consumerreview-8-the-growing-power-of-consumers.pdf. Accessed 12 May 2021.

113. Ibid.

114. Zheng, Jiajia, and Sangwon Suh. "Strategies to Reduce the Global Carbon Footprint of Plastics." Nature Climate Change, 15 April 2019, https://www.nature.com/articles/s41558-019-0459-z. Accessed 26 September 2021.

115. National Park Service. "How Long Will Litter Last?" 6 April 2015, https://www.nps.gov/olym/learn/kidsyouth/how-long-will-litter-last.htm. Accessed 5 October 2021; Root, Tik. "Why Carrying Your Own Fork and Spoon Helps Solve the Plastic Crisis." National Geographic, 28 June 2019, https://www.nationalgeographic.com/environment/article/carrying-your-own-fork-spoon-help-plastic-crisis. Accessed 11 October 2021.

116. Wen, Zongguo, Yiling Xie, Muhan Chen, and Christian Doh Dinga. "China's Plastic Import Ban Increases Prospects of Environmental Impact Mitigation of Plastic Waste Trade Flow Worldwide." Nature Communications, 18 January 2021, https://www.nature.com/articles/s41467-020-20741-9. Accessed 9 October 2021.

117. Geyer, R., J.R. Jambeck, and K.L. Law. "Production, Use, and Fate of All Plastics Ever Made." Science Advances, vol 3, no. 7, 2017, e1700782, http://advances.sciencemag.org/content/3/7/e1700782. Accessed July 21, 2022.

118. Australian Institute for Bioengineering and Nanotechnology (AIBN). "AIBN Nanotechnologist Turning Sugarcane Waste into Sustainable Packaging." 20 October 2020,

https://aibn.uq.edu.au/article/2020/10/aibn-nanotechnologist-turning-sugarcanewaste-sustainable-packaging. Accessed 3 April 2022.

119. Arnold, Christopher. "The Foundation for Economies Worldwide Is Small Business." International Federation of Accountants, June 26, 2019, https://www.ifac.org/knowledgegateway/contributing-global-economy/discussion/foundation-economies-worldwidesmall-business-0. Accessed 28 November 2021.

120. Science Based Targets initiative (SBTi). "Companies Taking Action." https://sciencebasedtargets.org/companies-taking-action?ambitionToggle=1#table. Accessed 7 March 2022.

121. Ritchie, Hannah. "Sector by sector: where do global greenhouse gas emissions come from?." Our World in Data, https://ourworldindata.org/ghg-emissions-by-sector. Accessed 29 May 2024

122. United Nations Industrial Development Organization (UNIDO). "Circular Economy." 2017, https://www.unido.org/sites/default/files/2017-07/Circular_Economy_UNIDO_0.pdf. Accessed 11 September 2021.

123. Krajewski, Markus. "The Great Lightbulb Conspiracy," IEEE Spectrum, 24 September 2014, https://spectrum.ieee.org/the-great-lightbulb-conspiracy. Accessed 25 September 2021.

124. Ibid.

125. Ibid.

126. Federal Trade Commission. Nixing the Fix: An FTC Report to Congress on Repair Restrictions. May 2021, https://www.ftc.gov/system/files/documents/reports/nixing-fix-ftc-reportcongress-repair-restrictions/nixing_the_fix_report_final_5521_630pm-508_002.pdf. Accessed 8 February 2022.

127. Department for Business, Energy & Industrial Strategy, The Rt Hon Kwasi Kwarteng MP, and Lord Callanan. "Electrical Appliances to be Cheaper to Run and Last Longer with New Standards." Climate Change and Energy, 10 March 2021, https://www.gov.uk/government/news/electrical-appliances-to-be-cheaper-to-run-and-last-longer-withnew-standards. Accessed 25 September 2021.

128. Carbon Trust. "2020 Consumer Research Shows Sustained Support for Carbon Labelling on Products." 23 April 2020, https://www.carbontrust.com/news-and-events/news/2020-consumer-research-shows-sustained-support-for-carbon-labelling-on. Accessed 19 September 2021.

129. Kateman, Brian. "Carbon Labels Are Finally Coming to the Food and Beverage Industry." Forbes, 20 July 2020, https://www.forbes.com/sites/briankateman/2020/07/20/

carbon-labels-are-finally-coming-to-the-food-and-beverage-industry/. Accessed 13 September 2021.

130. International Energy Agency (IEA). "The Oil and Gas Industry in Net Zero Transitions." November 2023, https://www.iea.org/reports/the-oil-and-gas-industry-in-net-zero-transitions. Accessed 29 May 2023.

131. The World Bank. "Services, Value Added (% of GDP)." 2019, https://data.worldbank.org/indicator/NV.SRV.TOTL.ZS. Accessed 26 November 2021.

132. Brogger, Tasneem Hanfi, and Alastair Marsh. "Big Banks Haven't Quit Fossil Fuel, With $4 Trillion Since Paris." Bloomberg, 25 October 2021, https://www.bloomberg.com/news/articles/2021-10-25/big-banks-haven-t-quit-fossil-fuel-with-4-trillion-since-paris. Accessed 26 November 2021.

133. United Nations Net-Zero Banking Alliance (NZBA). "Net-Zero Banking Alliance at COP 26." 5 November 2021, https://www.unepfi.org/news/industries/banking/net-zerobanking-alliance-at-cop-26/. Accessed 8 February 2022.

134. Kim, Suntae, Matthew J. Karlesky, Christopher G. Myers, and Todd Schifeling. "Why Companies Are Becoming B Corporations." Harvard Business Review, 17 June 2016, https://hbr.org/2016/06/why-companies-are-becoming-b-corporations. Accessed 27 November 2021.

135. Global Energy Monitor. "Summary Tables: Coal Plants by Country (Power Stations)." January 2024, https://globalenergymonitor.org/projects/global-coal-plant-tracker/summary-tables/. Accessed 29 May 2024; Hannah Ritchie and Pablo Rosado. "Electricity Mix." January 2024. Our World In Data, https://ourworldindata.org/electricity-mix. Accessed 29 May 2024; United States Energy Information Administration (EIA). "EIA projects nearly 50% increase in world energy usage by 2050, led by growth in Asia." 24 September 2019. https://www.eia.gov/todayinenergy/detail.php?id=41433. Accessed 12 July 2021.

136. International Energy Agency (IEA). "Net Zero by 2050." 2021, https://www.iea.org/reports/net-zero-by-2050. Accessed 11 July 2021.

137. United Nations. "Affordable and Clean Energy: Why it Matters." 2020, https://www.un.org/sustainabledevelopment/wp-content/uploads/2016/08/7_Why-It-Matters-2020.pdf. Accessed 13 February 2022.

138. Ritchie, Hannah, and Max Roser. "Electricity Mix." Our World in Data, 2020, https://ourworldindata.org/electricity-mix. Accessed 17 October 2021.

139. International Energy Agency (IEA), Electricity Market Report. February 2023, https://www.iea.org/reports/electricity-market-report-2023. Accessed 29 May 2024.

140. Global Energy Monitor. "Summary Data Coal Plants by Country (Power Stations)." July 2021, https://globalenergymonitor.org/projects/global-coal-plant-tracker/summarydata/. Accessed 17 October 2021; Global Energy Monitor. "Global Coal Plant Tracker." July 2021, https://globalenergymonitor.org/projects/global-coal-plant-tracker/tracker/. Accessed 17 October 2021.

141. "History of Power: The Evolution of the Electric Generation Industry." Power Magazine, 22 December 2020, https://www.powermag.com/history-of-power-the-evolution-of-theelectric-generation-industry/. Accessed 18 October 2021.

142. Water Science School. "A Coal-Fired Thermoelectric Power Plant." United States Geological Survey, 8 June 2018 https://www.usgs.gov/special-topic/water-science-school/science/a-coal-fired-thermoelectric-power-plant?qt-science_center_objects=0#qtscience_center_objects. Accessed 18 October 2021.

143. Hannah Ritchie and Pablo Rosado. "Electricity Mix." January 2024. Our World In Data, https://ourworldindata.org/electricity-mix. Accessed 29 May 2024

144. International Energy Agency (IEA). World Energy Investment 2021. 9 June 2021, https://www.iea.org/reports/world-energy-investment-2021/executive-summary. Accessed 17 October 2021.

145. United Nations. "Affordable and Clean Energy: Why it Matters." 2020, https://www.un.org/sustainabledevelopment/wp-content/uploads/2016/08/7_Why-It-Matters-2020.pdf. Accessed 13 February 2022.

146. IPCC. "Chapter 3." Special Report: Global Warming of 1.5°C: Impacts of 1.5°C Global Warming on Natural and Human Systems, 2018, https://www.ipcc.ch/sr15/chapter/chapter-3/. Accessed 12 May 2021.

147. Department of Economic and Social Affairs: Indigenous Peoples. "United Nations Declaration on the Rights of Indigenous Peoples," 2 October 2007, https://www.un.org/development/desa/indigenouspeoples/declaration-on-the-rights-of-indigenouspeoples.html. Accessed: 11 January 2022.

148. Brown, T.W., T. Bischof-Niemz, K. Blok, C. Breyer, H. Lund, and B.V. Mathieseng." Response to 'Burden of Proof: A Comprehensive Review of the Feasibility of 100% Renewable-Electricity Systems.'" Renewable and Sustainable Energy Reviews, September 2018, https://doi.org/10.1016/j.rser.2018.04.113, https://www.sciencedirect.com/science/article/pii/S1364032118303307. Accessed 21 July 2022.

149. Thornhill, James. "Two Years On, Elon Musk's Big Battery Bet Is Paying Off in Australia." Bloomberg, 28 February 2020, https://www.bloomberg.com/news/articles/2020-02-28/two-years-on-musk-s-big-battery-bet-is-paying-off-in-australia. Accessed 12 July 2021.

150. OECD/IEA. "Renewable Electricity Output (% of Total Electricity Output)." IEA Statistics, 2018, https://data.worldbank.org/indicator/EG.ELC.RNEW.ZS. Accessed 21 July 2022.

151. Climate Reality Project. "Follow the Leader: How 11 Countries Are Shifting to Renewable Energy." 3 February 2016, https://www.climaterealityproject.org/blog/followleader-how-11-countries-are-shifting-renewable-energy. Accessed 11 July 2021.

152. Jacobson, Mark Z., Mark A. Delucchi, Zack A.F. Bauer, Jingfan Wang, Eric Weiner, and Alexander S. Yachanin. "100% Clean and Renewable Wind, Water, and Sunlight All-Sector Energy Roadmaps for 139 Countries of the World." Joule, 6 September 2017, https://www.cell.com/joule/pdf/S2542-4351(17)30012-0.pdf. Accessed 12 July 2021.

153. Jacobson, M.Z., M.A. Delucchi, M.A. Cameron, S.J. Coughlin, C. Hay, I.P. Manogaran, Y. Shu, and A.-K. von Krauland, Impacts of Green New Deal energy plans on grid stability, costs, jobs, health, and climate in 143 countries, One Earth, 1, 449-463, doi:10.1016/j.oneear.2019.12.003, 2019

154. United States Energy Information Administration. "What are Ccf, Mcf, Btu, and therms? How Do I Convert Natural Gas Prices in Dollars per Ccf or Mcf to Dollars per Btu or therm?" 1 June 2021, https://opc.mo.gov/understanding-your-bill/understandingyour-gas-bill.html. Accessed 24 June 2022.

155. Background." Natural Gas, 20 September 2013, http://naturalgas.org/overview/background/. Accessed 10 June 2021.

156. United States EPA. "Understanding Global Warming Potentials." Updated 5 May 2022, https://www.epa.gov/ghgemissions/understanding-global-warming-potentials. Accessed 21 July 2022.

157. Balcombe, Paul, Kris Anderson, Jamie Speirs, Nigel Brandon, and Adam Hawkes. "The Natural Gas Supply Chain: The Importance of Methane and Carbon Dioxide Emissions," October 2016. ACS Sustainable Chemistry & Engineering, https://pubs.acs.org/doi/10.1021/acssuschemeng.6b00144. Accessed 27 June 2022.

158. Centers for Disease Control (CDC). "Carbon Monoxide (CO) Poisoning Prevention." 19 January 2021, https://www.cdc.gov/nceh/features/copoisoning/index.html. Accessed 12 July 2021.

159. United States EPA. "Nitrogen Dioxide's Impact on Indoor Air Quality." Updated 3 March 2022, https://www.epa.gov/indoor-air-quality-iaq/nitrogen-dioxides-impact-indoor-airquality. Accessed 21 July 2022; United States EPA. "Integrated Science Assessment (ISA) for Oxides of Nitrogen – Health Criteria (Final Report, January 2016)." https://cfpub.epa.gov/ncea/isa/recordisplay.cfm?deid=310879. Accessed 21 September 2021.

160. Center for Sustainable Systems. "Carbon Footprint Factsheet." University of Michigan, 2020, http://css.umich.edu/factsheets/carbon-footprint-factsheet. Accessed 15 July 2021.

161. Energy Networks Australia. "Working from Home: What $2.78 a Day Gets You." Energy Networks Australia, https://www.peopleenergy.com.au/docs/Working-fromhome-electricity-costs_FINAL-1.pdf. Accessed 16 July 2021.

162. Centers for Disease Control (CDC). "Carbon Monoxide (CO) Poisoning Prevention." 19 January 2021, https://www.cdc.gov/nceh/features/copoisoning/index.html. Accessed 12 July 2021.

163. Food and Agriculture Organization of the United Nations (FAO). "Climate Change Challenge Badge." 2015, http://www.fao.org/3/i5216e/i5216e.pdf. Accessed 1 August 2021.

164. Center for Sustainable Systems. "Carbon Footprint Factsheet." University of Michigan, 2020, http://css.umich.edu/factsheets/carbon-footprint-factsheet. Accessed 15 July 2021.

165. Deloitte Insight. "The Growing Power of Consumers." Deloitte, 2014, https://www2.deloitte.com/content/dam/Deloitte/uk/Documents/consumer-business/consumerreview-8-the-growing-power-of-consumers.pdf. Accessed 12 May 2021.

166. International Institute for Sustainable Development. "77 Countries, 100+ Cities Commit to Net Zero Carbon Emissions by 2050 at Climate Summit." 24 September 2019, http://sdg.iisd.org/news/77-countries-100-cities-commit-to-net-zero-carbon-emissionsby-2050-at-climate-summit/. Accessed 21 July 2021.

167. Global Energy Monitor. "Summary Data Coal Plants by Country (Power Stations)." Global Energy Monitor, July 2021, https://globalenergymonitor.org/projects/global-coalplant-tracker/summary-data/. Accessed 17 October 2021; Global Energy Monitor. "Global Coal Plant Tracker." Global Energy Monitor, July 2021, https://globalenergymonitor.org/projects/global-coal-plant-tracker/tracker/. Accessed 17 October 2021.

168. Ritchie, Hannah and Max Roser. "Electricity Mix." Our World in Data, 2020, https://ourworldindata.org/electricity-mix. Accessed 17 October 2021.

169. Grant, Don, David Zelinka, and Stefania Mitova. "Reducing CO2 Emissions by Targeting the World's Hyper-Polluting Power Plants." Environmental Research Letters, vol. 16, no. 9, 13 July 2021, https://iopscience.iop.org/article/10.1088/1748-9326/ac13f1. Accessed 31 July 2021.

170. Ibid.

171. Global Fuel Economy Initiative (GFEI). "GFEI re-launch document." May 2019. https://www.globalfueleconomy.org/data-and-research/publications/gfei-re-launch-document.

Accessed: 5 August 2021; International Transport Forum (ITF) and OECD. "ITF Transport Outlook 2021." 17 May 2021. https://www.itf-oecd.org/itf-transport-outlook-2021. Accessed: 26 February 2022.

172. The World Bank. "Urban Development." 20 April 2020, https://www.worldbank.org/en/topic/urbandevelopment/overview#1. Accessed 26 February 2022.

173. Ibid.

174. 1. Global Fuel Economy Initiative (GFEI), "GFEI Re-Launch Document." Global Fuel Economy Initiative (GFEI), May 2019, https://www.globalfueleconomy.org/data-andresearch/publications/gfei-re-launch-document. Accessed 5 August 2021; 2. and 3.International Transport Forum-OECD. "ITF Transport Outlook 2021." 17 May 2021, https://www.itf-oecd.org/itf-transport-outlook-2021. Accessed 26 February 2022.

175. International Transport Forum-OECD. "ITF Transport Outlook 2021." 17 May 2021, https://www.itf-oecd.org/itf-transport-outlook-2021. Accessed 26 February 2022.

176. Department of Energy. "Timeline: History of the Electric Car." https://www.energy.gov/timeline/timeline-history-electric-car. Accessed 5 March 2022.

177. International Energy Agency. "Global EV Outlook 2024." April 2024, www.iea.org/reports/global-ev-outlook-2024. Accessed 29 May 2024.

178. Rayner, Tim. "Taking the Slow Route to Decarbonisation? Developing Climate Governance for International Transport." Earth System Governance, June 2021, https://doi.org/10.1016/j.esg.2021.100100, https://www.sciencedirect.com/science/article/pii/S2589811621000045. Accessed 5 March 2022; Bullock, Simon, Alice Larkin, and James Mason. "Here's How Shipping Can Change Course to Hit Emissions Targets." World Economic Forum, 16 November 2021, https://www.weforum.org/agenda/2021/11/shipping-emissions-must-fall-a-third-by-2030-net-zero-2050/. Accessed 5 March 2022.

179. Ritchie, Hannah. "Which Form of Transport Has the Smallest Carbon Footprint?" 13 October 2020, Our World in Data, https://ourworldindata.org/travel-carbon-footprint. Accessed 27 February 2022.

180. ITF-OECD. "ITF Transport Outlook 2021." OECD International Transport Forum, May 2021, https://www.itf-oecd.org/itf-transport-outlook-2021. Accessed 4 August 2021.

181. Motavalli, Jim. "Every Automaker's EV Plans Through 2035 And Beyond." Forbes, 4 October 2021, https://www.forbes.com/wheels/news/automaker-ev-plans/. Accessed 24 October 2021

182. US Department of Energy. "Driving More Efficiently." United States Department of Energy – Office of Renewable Energy and Efficiency, https://www.fueleconomy.gov/feg/driveHabits.jsp. Accessed 26 October 2021.

183. Ibid.

184. Center for Sustainable Systems. "Carbon Footprint Factsheet." University of Michigan, 2020, http://css.umich.edu/factsheets/carbon-footprint-factsheet. Accessed 15 July 2021.

185. The World Health Organization. "Ambient air pollution data." 2021, https://www.who.int/data/gho/data/themes/air-pollution/ambient-air-pollution. Accessed 2 November 2021.

186. World Heart Federation. "Air Pollution and Cardiovascular Disease: A Window of Opportunity." 11 Jun 2019, https://world-heart-federation.org/news/air-pollution-andcardiovascular-disease-a-window-of-opportunity/. Accessed 21 July 2022.

187. American Heart Association. "American Heart Association Recommendations for Physical Activity in Adults and Kids." 2021, https://www.heart.org/en/healthy-living/fitness/fitness-basics/aha-recs-for-physical-activity-in-adults. Accessed 28 October 2021.

188. International Energy Agency (IEA). "Net Zero by 2050." 2021, https://www.iea.org/reports/net-zero-by-2050. Accessed 11 July 2021.

189. Ibid.

190. Ibid.

191. Ibid.

192. Ibid.

193. Khatib AN. Climate Change and Travel: Harmonizing to Abate Impact. Curr Infect Dis Rep. 2023;25(4):77-85. https://www.ncbi.nlm.nih.gov/pmc/articles/PMC9975868. Accessed: 29 May 2024

194. World Tourism Organization. "Vaccines and Reopen Borders Driving Tourism's Recovery." 4 October 2021, https://www.unwto.org/taxonomy/term/347. Accessed 28 October 2021.

195. World Tourism Organization. "International Tourism Highlights." 2019, https://www.e-unwto.org/doi/pdf/10.18111/9789284421152. Accessed 30 October 2021.

196. Schuurman, Richard. "Boeing Remains Cautious about Hydrogen." 27 July 2021, Air Insight Group, https://airinsight.com/boeing-remains-cautious-about-hydrogen/. Accessed 31 October 2021."

197. Ibid.

198. International Air Transport Association (IATA). "What is SAF?" https://www.iata.org/contentassets/d13875e9ed784f75bac90f000760e998/saf-what-is-saf.pdf. Accessed 31 October 2021.

199. "ASL Aviation Holdings Signs LOI with Universal Hydrogen, Becoming a Launch Customer for Hydrogen-Powered ATR 72 Cargo Aircraft." Business Wire, 12 October 2021, https://www.businesswire.com/news/home/20211011005712/en/ASL-Aviation-Holdings-Signs-LOI-with-Universal-Hydrogen-Becoming-a-Launch-Customer-for-Hydrogen-Powered-ATR-72-Cargo-Aircraft. Accessed 31 October 2021.

200. World Bank, "Climate Change," 8 April 2022, https://www.worldbank.org/en/topic/climatechange/overview#1. Accessed 10 April 2022; The World Bank. "Rapid, Climate-Informed Development Needed to Keep Climate Change from Pushing More than 100 Million People into Poverty by 2030." 8 November 2015. https://www.worldbank.org/en/news/feature/2015/11/08/rapid-climate-informed-development-needed-to-keepclimate-change-from-pushing-more-than-100-million-people-into-poverty-by-2030. Accessed 21 May 2022

201. United Nations. United Nations Framework Convention on Climate Change. 1992, https://unfccc.int/resource/docs/convkp/conveng.pdf. Accessed 12 March 2022.

202. Ritchie, Hannah. "Who Has Contributed Most to Global CO2 Emissions?" Our World in Data, 1 October 2019, https://ourworldindata.org/contributed-most-global-co2. Accessed 9 June 2021.

203. Buis, Alan. "The Atmosphere: Getting a Handle on Carbon Dioxide." NASA's Jet Propulsion Laboratory, 9 October 2019, https://climate.nasa.gov/news/2915/theatmosphere-getting-a-handle-on-carbon-dioxide/. Accessed 12 March 2022.

204. Kartha, Sivan, Eric Kemp-Benedict, Emily Ghosh, and Anisha Nazareth from the Stockholm Environment Institute, and Tim Gore from Oxfam. "The Carbon Inequality Era." September 2020, https://cdn.sei.org/wp-content/uploads/2020/09/researchreport-carbon-inequality-era.pdf. Accessed 14 November 2021.

205. United Nations, Department of Economic and Social Affairs, World Population Prospects Data Booklet, 2015, https://population.un.org/wpp/Publications/Files/WPP2015_DataBooklet.pdf. Accessed 16 October 2021.

206. Guo, Yuming, Shanshan Li, and Qi Zhao. "World's Largest Study of Global Climate Related Mortality Links 5 Million Deaths a Year to Abnormal Temperatures." 8 July 2021, Monash University, https://www.monash.edu/medicine/news/latest/2021-articles/worlds-largest-study-of-global-climate-related-mortality-links-5-million-deaths-a-yearto-abnormal-temperatures. Accessed 5 April 2022.

207. Intergovernmental Panel on Climate Change. "Chapter 29: Small Islands." Assessment Report 5, 2018, United Nations, https://www.ipcc.ch/site/assets/uploads/2018/02/WGIIAR5-Chap29_FINAL.pdf. Accessed 10 April 2022.

208. United Nations. "United Nations Framework Convention on Climate Change." 1992, https://unfccc.int/resource/docs/convkp/conveng.pdf. Accessed 12 March 2022.

209. United Nations Framework Convention (UNFCCC). "Green Climate Fund." https://unfccc.int/climatefinance?gcf_home. Accessed 13 March 2022.

210. United Nations. "COP26: Together for Our Planet." 2021, https://www.un.org/en/climatechange/cop26. Accessed 13 March 2022.

211. Green Climate Fund. "About GCF." https://www.greenclimate.fund/about. Accessed 13 March 2022.

212. United Nations. "World Economic Situation Prospects." 2020, Economic Analysis and Policy Division (EAPD) of the Department of Economic and Social Affairs of the United Nations Secretariat (UN DESA), https://www.un.org/development/desa/dpad/wpcontent/uploads/sites/45/WESP2020_Annex.pdf. Accessed 10 April 2022.

213. United Nations Development Programme. "Latest Human Development Index Ranking." 2020, United Nations, https://hdr.undp.org/en/content/latest-humandevelopment-index-ranking. Accessed 14 April 2020.

214. United Nations Development Programme. "Human Development Index (HDI)." 2020, United Nations, https://hdr.undp.org/en/content/human-development-index-hdi. Accessed 10 April 2022.

215. United Nations Committee for Development Policy Secretariat. "LDC Data." 2021, Department of Economic and Social Affairs Economic Analysis, https://www.un.org/development/desa/dpad/least-developed-country-category/ldc-data-retrieval.html. Accessed 10 April 2022.

216. United Nations Framework Convention on Climate Change. "Carbon Offset Platform: FAQ," https://offset.climateneutralnow.org/faq. Accessed 30 January 2022.

217. IPCC. "Glossary." Special Report: Global Warming of 1.5 °C: Impacts of 1.5°C Global Warming on Natural and Human Systems, 2018, https://www.ipcc.ch/sr15/chapter/glossary/. Accessed 12 May 2021.

218. United Nations Framework Convention on Climate Change. "Climate Neutral Now." https://unfccc.int/climate-action/climate-neutral-now. Accessed 30 January 2022.

219. IPCC. "Glossary." Special Report: Global Warming of 1.5 °C: Impacts of 1.5°C Global Warming on Natural and Human Systems, 2018, https://www.ipcc.ch/sr15/chapter/glossary/. Accessed 12 May 2021.

220. Gold Standard. "Carbon Offsetting Guide." 2020, https://www.goldstandard.org/sites/default/files/documents/gold_standard_offsetting_guide.pdf. Accessed 20 January 2022.

221. British Standards Institution (BSI). "Why is Net Zero Important for Businesses Including SMEs." https://www.bsigroup.com/en-GB/topics/sustainable-resilience/netzero/its-time-for-smes-to-step-up-to-the-net-zero-challenge3/. Accessed 22 July 2022.

222. United Nations Framework Convention on Climate Change. "Climate Neutral Now." https://unfccc.int/climate-action/climate-neutral-now. Accessed 30 January 2022.

223. Glaser, M.B. "CO2 – Greenhouse Effect." 12 November 1982, Exxon Environmental Affairs Program. http://www.climatefiles.com/exxonmobil/1982-memo-to-exxonmanagement-about-co2-greenhouse-effect/. Accessed 4 April 2021; Shell Greenhouse Effect Working Group. "The Greenhouse Effect." May 1988, http://www.climatefiles.com/shell/1988-shell-report-greenhouse/. Accessed 3 April 2021.

224. Ibid.

225. Shaw, Henry. "1981 Internal Exxon 'CO2 Position Statement.'" 15 May 1981, Exxon Research & Engineering Technology Feasibility Center, http://www.climatefiles.com/exxonmobil/co2-research-program/1981-internal-exxon-co2-position-statement/. Accessed: 3 April 2021.

226. American Petroleum Institute (API). "Draft Global Climate Science Communications Plan." 3 April 1998, http://www.climatefiles.com/trade-group/american-petroleum institute/1998-global-climate-science-communications-team-action-plan/. Accessed 22 July 2022.

227. Cushman, John H., Jr. "Industrial Group Plans to Battle Climate Treaty." 26 April 1998, The New York Times, https://www.nytimes.com/1998/04/26/us/industrial-groupplans-to-battle-climate-treaty.html. Accessed 4 April 2021.

228. Exxon. "ExxonMobil Pamphlet: Global Climate Change Everyone's Debate." October 1998, Climatefiles, http://www.climatefiles.com/exxonmobil/1998-exxon-pamphletglobal-climate-change-everyones-debate/. Accessed 16 May 2021.

229. Hiltzik, Michael. "A New Study Shows How Exxon Mobil Downplayed Climate Change When It Knew the Problem Was Real." 22 August 2017, Los Angeles Times, https://www.latimes.com/business/hiltzik/la-fi-hiltzik-exxonmobil-20170822-story.html. Accessed 5 April 2021.

230. CNN. "Kyoto Protocol Fast Facts." 18 March 2021, CNN Editorial Research, https://www.cnn.com/2013/07/26/world/kyoto-protocol-fast-facts/index.html. Accessed 5 April 2021.

231. UNFCCC. "What is the Kyoto Protocol?" UNFCCC, https://unfccc.int/kyoto_protocol. Accessed 1 June 2021.

232. Influence Map. "How the Oil Majors Have Spent $1Bn Since Paris on Narrative Capture and Lobbying on Climate." March 2019, InfluenceMap, https://influencemap.org/report/How-Big-Oil-Continues-to-Oppose-the-Paris-Agreement-38212275958aa21196dae3b76220bddc. Accessed 7 April 2021.

233. Kirk, Karin. "Fossil Fuel Political Giving Outdistances Renewables 13 to One." 6 January 2020, Yale Climate Connections, https://yaleclimateconnections.org/2020/01/fossil-fuelpolitical-giving-outdistances-renewables-13-to-one/. Accessed 7 April 2021.

234."Summons Jury Trial Demanded [City of New York v. ExxonMobil, BP, Royal Dutch Shell, and the American Petroleum Institute]." Supreme Court of the State of New York County of New York, 22 April 2021, https://www.eenews.net/assets/2021/04/23/document_cw_02.pdf. Accessed 14 June 2021.

235. Zdun, Matt. "Refugees of a Different Kind are Being Displaced by Rising Seas—and Governments Aren't Ready." 13 August 2017, CNBC News, 1 October 2019, https://www.cnbc.com/2017/08/11/climate-change-refugees-grapple-with-effects-of-rising-seas.html; Edmond, Charlotte. "5 Places Relocating People Because of Climate Change." 29 Jun 2017, World Economic Forum, 1 October 2019, https://www.weforum.org/agenda/2017/06/5-places-relocating-people-because-of-climate-change/; Esri Story Maps Team. "Climate Migrants." 2017, Esri, http://storymaps.esri.com/stories/2017/climate-migrants/index.html. Accessed 11 March 2021.

236. Kulp, S.A., and B.H. Strauss. "New Elevation Data Triple Estimates of Global Vulnerability to Sea-Level Rise and Coastal Flooding." Nature Communication, vol. 10, 29 October 2019, https://doi.org/10.1038/s41467-019-12808-z. Accessed 6 March 2021.

237. NASA. "World of Change: Global Temperatures." https://earthobservatory.nasa.gov/world-of-change/global-temperatures. Accessed 19 June 2022.

238. Lustgarten, Abrahm. "The Great Climate Migration." 23 July 2020, The New York Times, https://www.nytimes.com/interactive/2020/07/23/magazine/climate-migration.html. Accessed 15 March 2020.

239. "Migrant Crisis: Migration to Europe Explained in Seven Charts." 4 March 2016, BBC, https://www.bbc.com/news/world-europe-34131911. Accessed 12 May 2021.

240. Buchholz, Katharina. "Rising Sea Levels Will Threaten 200 Million People by 2100." 11 February 2020, https://www.statista.com/chart/19884/number-of-people-affected-byrising-sea-levels-per-country/

241. Islam, S. Nazrul, and John Winkel. "Climate Change and Social Inequality: DESA Working Paper No. 152 ST/ESA/2017/DWP/152." United Nations Department of Economics and Social Affairs, October 2017, https://www.un.org/esa/desa/papers/2017/wp152_2017.pdf. Accessed 2 January 2022.

242. Ibid.

243. World Health Organization (WHO). "Climate Change and Health." 30 October 2021, https://www.who.int/news-room/fact-sheets/detail/climate-change-and-health. Accessed 2 January 2022.

244. Ryan, Sadie J., Colin J. Carlson, Erin A. Mordecai, and Leah R. Johnson. "Global Expansion and Redistribution of Aedes-Borne Virus Transmission Risk with Climate Change." PLOS Neglected Tropical Diseases, 28 March 2019, https://journals.plos.org/plosntds/article?id=10.1371/journal.pntd.0007213.

245. Jordan, Rob. "How Does Climate Change Affect Disease?" March 15, 2019, Stanford Woods Institute for the Environment. https://earth.stanford.edu/news/how-doesclimate-change-affect-disease. Accessed 13 March 2021; Ryan, Sadie J., et al. "Global Expansion and Redistribution of Aedes-Borne Virus Transmission Risk with Climate Change." PLOS Neglected Tropical Diseases, 28 March 2019, https://journals.plos.org/plosntds/article?id=10.1371/journal.pntd.0007213. Accessed 2 January 2022.

246. Pagano, Anthony M., and Terrie M. Williams, "Physiological Consequences of Arctic Sea Ice Loss on Large Marine Carnivores: Unique Responses by Polar Bears and Narwhals." The Journal of Experimental Biology, vol. 224 (Suppl_1), February 2021, https://journals.biologists.com/jeb/article/224/Suppl_1/jeb228049/237178/Physiological-consequencesof-Arctic-sea-ice-loss. Accessed 4 January 2022.

247. Layton, Cayne, et al. "Kelp Forest Restoration in Australia." 14 February 2020, Frontiers in Marine Science, https://www.frontiersin.org/articles/10.3389/fmars.2020.00074/full. Accessed 17 April 2021; Australian Government Great Barrier Reef Marine Park Authority. "Climate Change." https://www.gbrmpa.gov.au/our-work/threats-to-the-reef/climate-change. Accessed 2 January 2022; IPCC. "Chapter 3." Special Report: Global Warming of 1.5 °C: Impacts of 1.5°C Global Warming on Natural and Human Systems. 2018, https://www.ipcc.ch/sr15/chapter/chapter-3/. Accessed 12 May 2021.

248. IPCC. "Chapter 3." IPCC Special Report: Global Warming of 1.5 °C.": Impacts of 1.5°C Global Warming on Natural and Human Systems. 2018, https://www.ipcc.ch/sr15/chapter/chapter-3/. Accessed 12 May 2021.

249. Ibid.

250. Eddy, Tyler D., et al., "Global Decline in Capacity of Coral Reefs to Provide Ecosystem Services." One Earth, 17 September 2021, https://doi.org/10.1016/j.oneear.2021.08.016. Accessed 21 September 2021.

251. US Federal Emergency Management Agency. "CPG 101: Developing and Maintaining Emergency Operations Plans." Federal Emergency Management Agency (United States), November 2010, https://www.fema.gov/pdf/plan/glo.pdf. Accessed 5 June 2021.

252. IPCC. "Managing the Risks of Extreme Events and Disasters to Advance Climate Change Adaptation." Intergovernmental Panel on Climate Change (IPCC), 2012, https://www.ipcc.ch/site/assets/uploads/2018/03/SREX_Full_Report-1.pdf. Accessed 5 June 2021.

253. BSI Standards Limited. "BS11200:2014 Crisis Management: Guidance and Good Practice." May 2014, BSI Standards Limited, https://shop.bsigroup.com/ProductDetail?pid=000000000030274343.

254. NSIDC. "Thermodynamics: Albedo." All About Sea Ice, 3 April 2020, National Snow & Ice Data Center (NSIDC), https://nsidc.org/cryosphere/seaice/processes/albedo.html. Accessed 3 March 2021.

255. Ibid.

256. NASA. "The Study of Earth as an Integrated System," NASA, https://climate.nasa.gov/nasa_science/science/. Accessed 3 March 2021.

257. Ibid.

258. Slater, Thomas, Isobel R. Lawrence, Inès N. Otosaka, Andrew Shepherd, Noel Gourmelen, Livia Jakob, Paul Tepes, Lin Gilbert, and Peter Nienow. "Review Article: Earth's Ice Imbalance." 25 Jan 2021, European Geosciences Union, https://tc.copernicus.org/articles/15/233/2021/. Accessed 4 March 2021.

259. Lindsey, Rebecca. "Climate Change: Global Sea Level." 14 August 2020, National Oceanic and Atmospheric Administration (NOAA), https://www.climate.gov/news-features/understanding-climate/climate-change-global-sea-level. Accessed 4 January 2022.

260. Turetsky, Merritt R., et al. "Permafrost Collapse is Accelerating Carbon Release." Nature, 30 April 2019, https://www.nature.com/articles/d41586-019-01313-4. Accessed 15 June 2021.

261. US Environmental Protection Agency. "Thawing Permafrost." 5 September 2017, https://archive.epa.gov/climatechange/kids/impacts/signs/permafrost.html. Accessed 15 June 2021.

262. IPCC. "Chapter 3." Special Report: Global Warming of 1.5 °C: Impacts of 1.5°C Global Warming on Natural and Human Systems, 2018, https://www.ipcc.ch/sr15/chapter/chapter-3/.Accessed 12 May 2021.

263. Smith, Esprit. "Human Activities Are Drying Out the Amazon: NASA Study." NASA's Earth Science News Team, 5 November 2019, https://climate.nasa.gov/news/2928/human-activities-are-drying-out-the-amazon-nasa-study/. Accessed 21 May 2022.

264. "US Environmental Protection Agency. "Thawing Permafrost." 5 September 2017, https://archive.epa.gov/climatechange/kids/impacts/signs/permafrost.html. Accessed 15 June 2021.

265. Smith, Esprit. "Human Activities Are Drying Out the Amazon: NASA Study." NASA's Earth Science News Team, 5 November 2019, https://climate.nasa.gov/news/2928/human-activities-are-drying-out-the-amazon-nasa-study/. Accessed 21 May 2022.

266. Gunders, Dana, Jonathan Bloom, JoAnne Berkenkamp, Darby Hoover, Andrea Spacht, and Marie Mourad. "Wasted: How America is Losing Up to 40 Percent of Its Food from Farm to Fork to Landfill." The Natural Resources Defense Council (NRDC), August 2017, https://www.nrdc.org/sites/default/files/wasted-2017-report.pdf. Accessed 22 July 2022.

267. IPSOS. "Garot Law: Where is the Application of Anti-Waste Measures?" 20 February 2018, https://www.ipsos.com/fr-fr/loi-garot-ou-en-est-lapplication-des-mesures-antiga spillage. Accessed 14 September 2021.

268. Ibid.

269. UN Environment Programme. "Food Waste Index Report 2021." 2021, https://www.unep.org/resources/report/unep-food-waste-index-report-2021. Accessed 6 November 2021.

270. Ibid.

271. IPCC. "Chapter 5: Food Security." Special Report: Climate Change and Land, January 2020, https://www.ipcc.ch/srccl/. Accessed 13 August 2021.

272. Purdue University "PICS: How a technology scaled up to reach millions." 13 September 2018. https://ag.purdue.edu/stories/pics-how-a-technology-scaled-up-toreach-millions/. Accessed: 8 August 2022.

273. Kitinoja, Lisa. "Innovative Technologies for Reducing Food Waste." 5 August 2016, Huffington Post, https://www.huffpost.com/entry/innovative-technologies-f_b_11356568. Accessed 20 November 2021.

274. Australian Cane Farmers Association. "Sugarcane Waste-Based Durable Packaging is Plastic-Free and Compostable." 29 September 2020, https://www.acfa.com.au/category/bioplastics/. Accessed 7 February 2022.

275. US Department of Agriculture – Agricultural Research Service. "What is Pyrolysis?" 10 September 2021, https://www.ars.usda.gov/northeast-area/wyndmoor-pa/eastern regional-research-center/docs/biomass-pyrolysis-research-1/what-is-pyrolysis/. Accessed 21 November 2021.

276. Ibid.

277. Ibid.

278. Campbell, Lindsay. "Going Green: Can Electric Tractors Overtake Diesel?" Modern Farmer, 28 March 2020, https://modernfarmer.com/2020/03/going-green-can-electrictractors-override-diesel/. Accessed 29 November 2021.

279. Beyond Zero Emissions. "Electrifying Industry 2020." Beyond Zero Emissions, 2020, https://bze.org.au/wp-content/uploads/2020/12/electrifying-industry-bze-report-2018.pdf. Accessed 23 August 2021.

280. Ibid.

281. Collins, Leigh. "'World First' as Hydrogen Used to Power Commercial Steel Production." Recharge News, 28 April 2020, https://www.rechargenews.com/transition/-world-first-as-hydrogen-used-to-power-commercial-steel-production/2-1-799308. Accessed 21 September 2021.

282. International Energy Agency. "Iron and Steel Technology Roadmap." International Energy Agency, October 2020, https://www.iea.org/reports/iron-and-steel-technologyroadmap.Accessed 21 September 2021.

283. United Nations - Department of Economic and Social Affairs. "68% of the World Population Projected to Live in Urban Areas by 2050, Says UN." 16 May 2018, https://www.un.org/development/desa/en/news/population/2018-revision-of-world-urbanizationprospects.html. Accessed 27 October 2021.

284. United Nations Statistics Division. "The Sustainable Development Goals Report 2019." 2019, https://unstats.un.org/sdgs/report/2019/The-Sustainable-Development-Goals-Report-2019.pdf. Accessed 27 October 2021.

Made in United States
Troutdale, OR
11/26/2024